[英国] 尼克·米德尔顿 著　朱庆云 译

牛津通识读本· ————————

河流

Rivers

A Very Short Introduction

译林出版社

图书在版编目（CIP）数据

河流 ／（英）尼克·米德尔顿（Nick Middleton）著；
朱庆云译. —南京：译林出版社，2023.3
（牛津通识读本）
书名原文：Rivers: A Very Short Introduction
ISBN 978-7-5447-9407-7

Ⅰ.①河… Ⅱ.①尼… ②朱… Ⅲ.①河流－普及读
物 Ⅳ.①P941.77-49

中国版本图书馆 CIP 数据核字（2022）第 170518 号

著作权合同登记号　图字：10-2018-429 号

河流　[英国] 尼克·米德尔顿 ／著　朱庆云 ／译

责任编辑　许　丹
装帧设计　景秋萍
校　对　戴小娥
责任印制　董　虎

原文出版　Oxford University Press, 2012
出版发行　译林出版社
地　址　南京市湖南路 1 号 A 楼
邮　箱　yilin@yilin.com
网　址　www.yilin.com
市场热线　025-86633278
排　版　南京展望文化发展有限公司
印　刷　江苏扬中印刷有限公司
开　本　890 毫米 ×1260 毫米　1/32
印　张　8.25
插　页　4
版　次　2023 年 3 月第 1 版
印　次　2023 年 3 月第 1 次印刷
书　号　ISBN 978-7-5447-9407-7
定　价　39.00 元

序　言

张建云

　　河流,是人类文明的摇篮。无论是距今3 000～5 300年孕育于南亚次大陆印度河流域的哈拉帕文明,还是距今6 000多年诞生于底格里斯河和幼发拉底河流域的苏美尔文明,抑或是距今5 000多年发源于中国黄河流域和长江流域的华夏文明,以及距今7 000多年肇始于尼罗河的古埃及文明,无不与大江大河直接相关。

　　河流,是流动的赞美诗。刘禹锡用"九曲黄河万里沙,浪淘风簸自天涯"描述黄河从遥远的地方蜿蜒奔腾而来,一路裹挟着万里黄沙的景象,写出了黄河的地理地貌及河流特征。李白以"孤帆远影碧空尽,唯见长江天际流"的诗句,写出了长江万壑争流、千岩竞秀的磅礴之势,又将诗人的一片情意托付江水,将情与景完全交融。更有一首《长江之歌》,"我们赞美长江,你是无穷的源泉;我们依恋你,你有母亲的情怀",道出了人与河流的渊源与和谐。

　　河流,水也。上善若水,水善利万物而不争。是长江黄河哺育了中华文明，也是长江黄河支撑了中华民族的发展和复兴，更是长江黄河成就了我们炎黄子孙。对河流的了解，对河流的热爱，对河流的保护，其意义和影响深远。

　　我从1978年就读华东水利学院（现河海大学）陆地水文专业开始，就与河流结下了不解之缘，至今已有40多年。我曾经担任国际水文科学协会中国委员会主席十多年，对于《河流》一书的作者尼克·米德尔顿也算得上熟悉。我也曾经在水利部担任水文局总工和副局长长达10年，而本书中文版译者朱庆云高工一直供职于江苏省水文水资源勘测局南京分局，算是水文同仁。此次《河流》中文版出版，译林出版社邀我为之作序，而普及水情教育是一名科技工作者义不容辞的社会责任，我就欣然接受了这一任务。

　　本书作者尼克·米德尔顿算得上是一位奇人，他有着一连串令人炫目的头衔。首先，他是一位地理学家，游历过70多个国家，去过世界上最热、最冷、最湿和最干燥的有人居住的地方；他也是一名作家，目前已经出版了地理学相关专著16部，其著作被翻译成十几种语言，受到世界各地读者广泛欢迎；他还是一名电视节目主持人，撰写并主持的名为《极远之地》的电视纪录片，分别由英国第四频道和世界其他地区的国家地理频道播出；另外，尼克还在牛津大学教授地理学，是牛津大学圣安妮学院的研究员；最后，他曾作为顾问为联合国开发计划署、联合国环境规划署以及世界自然保护联盟、英国政府国际发展部提供环境问题咨询。相信这样一位奇人撰写的《河流》一书，会给读者带来丰富的科学知识和渊博的哲理。

　　《河流》一书大致可以分为两个板块：第一章和第五章为一

个板块，偏重于科学维度；第二、三、四章为另一个板块，偏重于人文维度。作者这样的结构安排，颇具匠心。无论是人文背景的读者，还是理工科背景的读者，都可以在书中找到自己的共鸣点。同时，扑面而来的新奇内容，也能让读者在阅读之后获得满满的收获感。

信息量巨大，是本书的一个显著特点。第一章作者对河流等级、河流类型、河流长度、河道流量、河岸侵蚀、河流生态、河流洪水等基本概念做了详细阐述，内容详尽，但文字浅显流畅，易于理解。第二章从神话中的河流、洪水传说、圣河、神圣的河流生物、河流中的精灵等方面讲起，为读者介绍了与河流相关的各种习俗、神话和宗教仪式。第三章用最初的文明、探索的通道、天然的屏障、河权与冲突、贸易与运输、欧洲大动脉几个小节，介绍了河流在不同方面对人类历史进程产生的影响。第四章从语言、绘画、音乐、文学、电影等方面介绍了河流在文化的诸多方面发挥的重要作用。第五章从灌溉农业、河流整治、土地利用、城市河流、全球变暖、河流恢复等方面着手，介绍了人类滥用河流带来的种种恶果以及人们对此进行的反思。本书的内容无疑会对我国生态文明建设、江河保护与绿色发展，以及流域高质量发展起到重要的借鉴作用。

合理开发利用和保护河流，是人类从文明诞生之初就面临的永恒主题。在这方面，中国人走出了一条独具特色的道路。中国传统哲学的一元论，强调道法自然、天人合一的思想。在改造自然的过程中，中国人一直遵循顺应自然而非对抗自然的精神。古代大禹通过疏导之法治理河流的故事，在中国家喻户晓。春秋战国时期秦昭王末年蜀郡太守李冰主持建造的都江堰堪称古代水利工程的经典之作，建堰2 000多年来经久不衰，而且发挥的效益愈来愈大。2000年，它被联合国教科文组织列入了世界文化遗产名录。同为世界文化遗产的大运河、良渚遗址中的水利工程体系，都是中国古代先民创造的河流利用的奇迹。然而，人类不合理的经济开发，导致河流泛滥，也造成了巨大的生命财产损失，教训极为惨痛。

保护环境，保护河流，是人类的共同责任。习近平总书记指出，绿水青山就是金山银山。当前实施的长江大保护战略，就是保护河流的具体体现。《河流》一书在此时出版，也为我们提供了一本水情教育的优秀读本。加强水情教育，强化河流保护意识，任重道远，这也是我愿意为本书作序的出发点之一，愿以此与各位读者共勉。

河流

2022年11月于南京

本书献给彻丽

目 录

引 言

从广袤的大陆到几乎最小的岛屿，都有河流在流动。从涓涓细流到壮阔的波涛，其形式几乎令人眼花缭乱。作为水的来源，河流一直是人们赞美的对象，也是人们关注的现实问题。它们是文明的摇篮，也是灾难的元凶。河流可以是屏障，也可以是通道；可以承载贸易和泥沙，也可以承载文化和冲突；可以激发灵感，也可以带来恐惧。

本书揭示了河流在我们这个星球及其居民的生活中所发挥的种种作用，强调了河流在从公共卫生到鱼类学，神学到文学批评等显性和隐性方面的重要性。河流的流动激发了诗人、画家、哲学家、科学家、探险家和朝圣者的灵感。不了解泰晤士河，就无法理解伦敦；不了解尼罗河，就无法理解埃及。河流赋予国家以国名，也决定了战争的胜负。

河流可以劈开深谷，如巨蛇一般蜿蜒穿过平原；可以从巨大的悬崖上跃下，如土地的手指一般伸入大海。河流主宰着景观，侵蚀并创造景观。毫无疑问，它们是一系列复杂的自然过程的产物。但是许多河流的演变受社会系统和自然系统的双重驱动，尽管乍一看这似乎令人惊讶。

在物质层面上，人类长期以来一直在与河流互动，从河流中取水和捕鱼，为满足自身需求而改造河流。反过来，河流又影响了历代文化的无数方面，催生了神话，且产生水力。河流在神话，宗教，以及诸如音乐、艺术和诗歌等社会的其他许多方面都占有一席之地。因此，它们不仅是作为物质世界一部分的物理实体，也是与社会系统相互作用的文化实体。在许多方面，河流像传输水一样传播价值。

因此，精确地定义河流并非易事，了解这一点并不令人意外。我们的朋友《牛津英语词典》认为，河流是"通过渠道流向大海或湖泊等的大量自然水流"。这一定义适用于许多河流，但并不适用于所有河流。在极寒之地，河流并非一直在流动。沙漠中的大多数河流也是如此。前一种情况下，河水长时间处于冰冻状态；而在后一种情况下，河道中往往根本没有水。"大量"这个词也有问题。许多读者会在大脑中区分河流和较小的水体，如小溪或小河。然而，并非所有人都进行这样的区分。在法律术语中，"河流"一词通常包括所有的自然水流，无论其大小。问题更大的是"自然"这个词。几千年来，人们一直在与河流互动并改造它们，因此，今天可以被描述为完全自然和未经改造的河流并不多。为了完整地解构字典定义，请记住，并非所有的河流都流入大海或湖泊。有些河流在到达另一个水体之前就消失在地下或完全干涸。

因此，与我们对字典的期望相反，字典的定义并不具有普遍性。这不是一个现代难题。19世纪末，詹姆士·克莱德博士

图1 河流在塑造景观中发挥了主要作用，有时以一种惊人的方式，如此处尼泊尔境内的喜马拉雅山脉   v

3

的《基础地理学》一书再版至第25版。1885年他在《苏格兰地理杂志》上发表的一篇关于"江河"的论文中曾经遇到同样的难题。但他回避了这个问题，通过转述约翰·斯图尔特·密尔的观点而放弃了定义的形而上的准确性，他说："对于河流的含义，每个人都有自己的概念，对于一般目的而言这些概念足够正确。"

　　一系列关于儿童如何感知周围世界的地理学研究表明，孩子对河流这一概念的认知毫不逊色。这里摘录了近期一项研究中孩子关于"河流是什么"的几个最佳答案："向下流淌的湿漉漉的水"，"湿漉漉的蓝色长条物"以及"流动的、有鱼和水的东西"。对于我们的目的而言，所有这些说法都足够正确。

大自然的推动力

水是大自然的推动力。

莱昂纳多·达·芬奇（1452—1519）

（意大利画家、建筑师和工程师）

我们生活在一颗湿润的星球上。水是地球上最丰富的物质，覆盖地表的三分之二。同时，空气、动植物以及大地中也存在少量的水。这些水处于持续运动之中，在陆地、海洋和大气中循环，这种不间断的更替被称为水文循环。河流在水文循环中发挥着关键作用，排走陆地上的水并最终将其送入海洋。

在重力的作用下，降雨或者融雪中未被蒸发或下渗的部分在地表上向低处流动。在细微不规则地形的引导下，水流汇成小溪，小溪合并为冲沟，之后再流进更大的河道。壤中流和地下蓄水增加了河流流量，但是，河流并不仅仅是流向大海的水。河流还挟带着岩石和其他沉积物、溶解矿物质、植物和死去或活着的动物。在此过程中，河流输送了大量的物质，为各种野生生物

1 提供了生境。河流的冲击和沉积作用催生了峡谷和冲积平原，在很大程度上塑造了地球上的大陆景观。

从河源到河口，整条河流处于渐次变化之中，狭窄湍急的陡峭河流逐渐变得宽阔、深邃且常常蜿蜒曲折。从上游到下游，河流发生着连续不断的变化，水量通常不断增加，泥沙粒径由粗变细。在上游，河流侵蚀河床和河岸，搬走土壤、卵石甚至巨石，而到了下游，则发生沉积。在河流物理特征发生变化的同时，以河流为家园的动物和植物的类型也在相应地变化。

其狭长的线性形式以及仅在一个方向上流动的特征，为如何描述和理解河流的物理、化学和生物特性提供了明显的空间维度：水平方向上，从上游到下游。但是，河流并不仅仅是一条河道，它还是所流经区域整体的一个组成部分，因此，将周围景观作为河流的横向维度，也是合适的。河流与景观（或者说河流景观，有些人更愿意这么说）之间的联系千丝万缕。这种联系既包括河流中大部分水经周围地形汇入河道这样的简单事实，也包括鲑鱼在河流中的重要性（即它是当地熊的季节性食物）。

第三个维度是垂向的。河流与河道下方的泥沙以及上方的大气相互作用。对于许多河流而言，水的来源有两种：一种直接来源于大气中的降雨——或者另一种降水形式（降雪）；另一种来自赋存于下方岩石及砾石孔隙中的地下水，两者都参与水文循环。

第四个至关重要的维度是时间，它在河流研究中也占有重要地位。这是因为影响河流的许多因素发生着巨大的变化，尤2 其是河道中的水量。从历时一小时以下的强暴雨到持续数百万年的构造力影响，这种变化在时间尺度上差异很大。

河流遍布世界各地，几乎在每一种景观上都留下了印迹。某些地区缺乏地表排水系统，但其中一些地区的河流位于地表以下。在沙漠中，许多河流在一年中的大部分时间里都是干涸的，只有偶尔发生暴雨时才会有水。在另外一些地方，化石河床和峡谷表明，在非常遥远的过去的某个时段内，这里曾经是河流。这种化石特征也存在于其他行星上：在火星和土星最大的卫星土卫六上已经发现了河床和峡谷，而且这些网状结构与地球上的河流和溪流的特征非常相似。火星表面上的这些特征是由昔日水流冲击而成的，但是土卫六上的河床和排水网络被认为是由液态甲烷流动形成的。对于地球上大部分陆地表面而言，流动的河流是最基本的元素之一。借由阳光和重力提供的能量，河流塑造了峡谷和坡地，为生物群落提供了复杂的生境。

河流等级

河流的一个有趣的方面是它们似乎是按等级构成的。从飞机上或地图上看，河流形成明显的树枝状网络。小的支流交汇后形成较大的河流，较大的河流交汇后形成更大的河流。通常采用数值分级方案来描述河流规模的渐次增加，其中最小的河流为一级河流，两条一级河流交汇后形成二级河流，两条二级河流交汇后形成三级河流，以此类推。只有两条级别相同的河流交汇，河流的级别才会增加。尼罗河和密西西比河等很大的河流为十级河流，亚马孙河为十二级河流。

河流的排水区域与河流的规模成正比。这一区域有几个不同的名称：流域、河流流域或者集水区（在美式英语中也用watershed来表述这样的区域，但在英式英语中这个词表示相邻

3

两个流域之间的分水岭）。正如水系由嵌套在高级别河流中的低级别河流构成一样，流域也相应地嵌合在一起形成嵌套等级结构。换句话说，较小的单元是嵌套在较大单元内的重复元素。所有这些单元通过水流、泥沙和能量联系在一起。

将河流看成由一系列等级单元构成，这种认识方式为研究与河流相关的模式和过程提供了一个有效框架。在最大尺度上，可以研究整个流域。在流域内，在渐次变小的尺度上，研究人员可以重点关注支流之间的某个特定河段、河段内的局部河段，以此类推直到河床上的一小块沙粒地。同时，这种分级方法强调，在较高等级上发生的过程会对较低等级的特征产生相当大的影响，但反之不然。在流域尺度上，重要影响因素包括气候、地质、植被和地形。在所有较小的尺度上，小到一小块沙粒地，这些因素都会产生影响。这样的小块沙粒地同时还承受其他局部影响，如流水中的涟漪，但水流的这些微小变化对整个流域的影响微乎其微。

与相关空间尺度对应，存在匹配的时间尺度，而这些空间尺度也可以分级。一般来说，空间尺度越大，过程越慢，变化率越小。例如，气候变化和地质变化的发生，其时间尺度很大，可达数百年至数百万年。而水中的涟漪，其发生的时间尺度则极小，只有几毫秒到几秒。

同样重要的是，我们不要忘记，大致而言，随着规模的增大，影响景观以及流经景观的河流的因素的复杂性也随之增加。例如一级河道的小集水区很可能只存在一种岩石类型，且处于一个气候区内。集水区越大，其拥有的岩石类型和跨越的气候区就越多，因而愈加复杂。

河流

4

河流类型

前面详细介绍的数值分级方案,是众多河流分类尝试中的一种。当我们把关注范围由河道扩展到整个流域时,就会出现无数种不同类型的河流或"河"系(fluvial,源自拉丁词 *fluvius*,意即"河流")。每一种河流分类都取决于研究者的视角,即研究者认为的最重要的方面。生物学家主要关注的可能是鱼或者水生植物等特定生物种群的分布。例如,不同的物种可能与不同的地形和地质类型相关联,因此,河流可能被归类为"山地"、"高地"、"低地白垩"、"低地砂岩"以及"低地和高地黏土"等。另一些研究者使用选定的化学因素作为分类的基础。例如使用pH值,于是河流可以分类为强酸性、弱酸性或碱性。自然保护相关部门可能结合所有这些甚至更多的观点。英国自然保护委员会根据植物群落对英格兰、威尔士和苏格兰的河流进行了分类,确定了4个主要河流群、10个类型和38个子类型。

另一种简单的方法是根据规模进行河流分类。一些权威机构更喜欢用"溪流"(stream)来表述规模谱系一端的河流。大江或者大河(这两个词都常用来表示规模谱系的另一端)通常是指流域大、河道长、输沙量大或者流量大的河流。我们已经注意到,河流长度和流域面积之间存在着一致性关系,但是,由于流域地质、地形和水文情势的变化,其他变量之间并不存在这种关系。当被要求说出世界上最大的河流时,大多数人可能都会列出一个类似前10条或前20条河流的名单,但要给出一个完美的定义却依然很困难。

依据水系在景观中形成的型式来区分不同类型的水系,是

5

一种常见方法。树状水系型式存在几种常见的变化形式，可使用不同的术语描述，包括树枝状、放射状、网格状、平行式和矩形。景观的地质状况是水系型式的主要影响因素。

依据水流类型对不同类型的河流进行分类，是一种简易的分类方法。终年连续流水的河道被称为"常年性"河流，但这一术语绝不能用来描述所有河流。有些河道只在特定季节才有流水。这些季节性或者"间歇性"河流可能存在于严冬时河水完全结冰，或者具有明显雨季的地区。具有更少常年水流的河流被称为"暂时性"河流，其组成河道仅在单次暴雨后数小时或数天内存在流水。在沙漠中出现和流动的河流是典型的暂时性河流。第四类是"间断性"河流，这种河流只在很短的河段内终年

图2　西伯利亚中部高原的卫星图像，这是典型水系的一个示例。高海拔
6　地区的积雪与无雪山谷形成对比，进一步突显了水系型式

有水，大部分河段是干涸的。就像自然界中的大多数分类方案一样，这些区别无疑是真实的，但是，我们最好将不同类别之间的边界看成水流情势类型连续体上的点。这是因为，例如，在持续数年的绵长多雨期内，暂时性河流可能呈现季节性河流的特征；而在干旱期内，季节性河流雨季时的流量可能为零或更具间歇性，使之更像暂时性河流。

河流有多长？

测量河流长度听起来似乎不是难事，但实际上要复杂得多。由于各种因素的制约，世界各地河流长度的测量和估算存在很大差异，这些因素包括季节、制图者的能力、所使用设备的质量以及测量内容的决策。从理论上讲，这项工作应该不复杂：确定河源与河口的位置，然后准确测量两者之间的河流长度。河口的位置通常很明确。一般来说，河流中心线与河流出口两侧连线的交叉点，就是河口的准确位置。

而河源准确位置的确定，常常要复杂得多。几个世纪以来，在偏远和人迹罕至的地区寻找特定河流的源头激发了探险家们的兴趣并激励着他们，时至今日仍然如此。

关于许多河流的真正河源，一直存在分歧，这也成了探险史一直都有的一个特征。从某种意义上说，寻找河流的"正"源注定是一项带有推测性的任务，因为大多数河流通常有许多支流，因此也就存在许多河源。对于大多数权威机构来说，距离河口最远的河源才是"正"源，这样可获得河流的最大长度。但是，毫不意外的是，在最远河源这一问题上也出现了分歧。

另一个使问题复杂化的因素是，要不要将那些具有不同河名的支流包括在内？在实践中，是否包含支流的系列决策可能是寻找河源任务的主要组成部分，而这些决策也是导致特定河流长度测量结果不一致的一个主要原因。以湄公河为例。所有人都承认湄公河发源于青藏高原，但河源的确切位置还存在争议。可能的源头包括果宗木查山、拉赛贡玛山、扎那日根山、查加日玛山和吉富山上的冰川。可能的其他河源还包括鲁布萨山口、伦格磨山口和扎西气娃湖。考虑到认定的河源的数量，对于湄公河被冠以世界第九长和第十二长河流的不同名头，而在世界其他大河河源确定过程中并未出现类似的混乱，也许就不会觉得奇怪了。大量文献记载表明，湄公河的长度在4 180～4 909千米之间。如果我们认可这条河流的正源在吉富山（当然很多人并不认同这一观点），那么沿着4 909千米的长度，这条河流有六个名字。在吉富山的两侧，融化的冰雪形成一条溪流，名为谷涌-高地扑溪（只有夏天有水）。20多千米后，这条河变成了郭涌曲河，之后，郭涌曲河变成扎阿曲河。扎阿曲河与扎那曲河交汇后形成扎曲河，扎曲河之后变成澜沧江，直到自中缅边界出境后改称湄公河，然后一直延伸到越南南部的三角洲。湄公河在三角洲分成几条支流，最终注入南海。

有人认为，这就是整条湄公河，长度为4 909千米。另外一些人认可4 909千米这一长度，但是他们认为，严格地讲，这条河的名称应该是湄公-澜沧-扎曲-扎阿曲-郭涌曲-谷涌-高地扑。还有一些人，他们更愿意只讨论被称为湄公河的河段，这种情况下，湄公河的长度实际上才2 711千米。另外一个阵营的观点差异更大，他们根本就不承认这条河的正源在吉富山。

如果你感到有些茫然，这是可以理解的。不过，还有更加让人困惑的，有些河流没有河口。南部非洲的奥卡万戈河在流入内陆的奥卡万戈三角洲沿途逐渐变小，三角洲的规模随季节变化而变化。因此，河流的确切终点也随季节而变化。有些河流不止一条河道。在"辫状"（见下文）河段中，应测量哪条河道的长度？施测时间也很重要。吉富山中的谷涌-高地扑溪只有在夏日融化季节才有水。并非终年有水的河道，应该计算在内吗？另一个有关测量时间的难题出现在季节性泛滥的河流上。例如，亚马孙流域的大片地区雨季时会发生洪水，旱季时仅沿弯道流动的水流此时会直接"漫地"流动。弯道的长度要不要计算在内？经较长一段时间后，河流会产生新的土地，例如，通过将泥沙沉积在三角洲处，使其长度增加。

河流长度测量的另一个重要问题是测量的比例尺。从根本上说，河流的长度随着地图比例尺的变化而变化，因为比例尺不同，细部概化程度也不同。河道沿线的地形非常复杂，细部地形常常相互嵌套。这种几何复杂性被称为"分形"，它是许多自然事物固有的一种特性，而这种复杂性有时会走向荒谬。但是，对更多细节的渴求在何种情况下会跨越界限而进入荒谬的领域呢？

利用卫星测绘技术和全球定位系统（GPS）确定河源的准确位置，将不断提高我们研究水系整体的能力。但是，关于研究规模以及包括和排除哪些支流的主观决策意味着，实际上我们依然不能准确确定哪条河流拥有"世界最长"的称号。几个世纪以来，随着知识的增长和标准的改变，人们始终无法确定亚马孙河和尼罗河哪一条为世界最长河流。1858年，苏格兰探险家

9

约翰·汉宁·斯佩克宣称发现维多利亚湖是尼罗河的源头，认为自己揭开了19世纪世界地理学的一个重大谜团。在20世纪的大部分时间里，大多数权威机构都认可尼罗河是世界最长河流，他们将从南部流入维多利亚湖的尼罗河最长支流包括了进去。然而，1990年代以来，人们在秘鲁南部山区进行了一系列的亚马孙河河源考察，之后，一些可信的主张表明，亚马孙河更长。这些主张认定亚马孙河的长度为约6 850千米，比尼罗河至少长150千米，但是争议不太可能就此结束。

河道流量

河道水流的两个特别重要的属性是流速和流量，流量（discharge）是指单位时间内通过特定河道断面[①]的水量，但令人不解的是，英语中也会直接用flow这个词表示流量。依据连续流量资料与时间的关系绘制的曲线被称为流量过程线，根据选

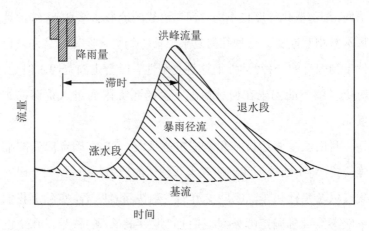

图3　显示河流流量随降雨过程变化的暴雨洪水过程线

择的时间坐标,流量过程线可以详细描述历时数天的洪水事件,
或者一年、一年以上的流量过程。

施测河道流量以及分析流量资料,对于水资源评价和水旱灾害评估非常重要。拥有全世界最悠久系列水文资料的河流非尼罗河莫属,公元641年,开罗的罗达岛上修建了尼罗河水位观测设施。作为负责管理罗达"水尺"的官员,埃尔·米哈斯长老的职责是观测水位并在洪水期通过公告员发布洪水每日上涨情况。这是埃及一年中最紧张的时候。如果河水没有达到一定的水位,许多农田将无水灌溉,饥荒也将在所难免;而在一定的水位之上,灌溉用水可以确保,政府的税收也就有了保证。埃尔·米哈斯长老的这一职位延续了1 000多年。该职位的最后一位任职者于1947年去世。1950年代,埃及政府决定在尼罗河的阿斯旺修建一座大坝,从而极大地改变了埃及与尼罗河的密切关系。阿斯旺大坝1970年最后竣工(见第五章),在计算大坝所需库容时,利用罗达水尺观测的水位资料发挥了巨大作用。

河道流量取决于许多不同因素,包括流域的面积和形状。如果其他条件相同,则流域越大,流量越大。与狭长型流域河道相比,圆形流域河道的洪峰流量往往更大,因为圆形流域中各支流汇流时间大体上相同,而狭长型流域中各支流则以交错的方式汇流。流域下垫面也是一个重要因素。例如,植被拦截降雨,从而延缓了其汇入河流的过程。

决定河道流量的一个特别重要的因素是气候。它是控制前述不同类型河流,即常年性、间歇性、暂时性和间断性河流的主要因素。流量最大的河流几乎全部位于全年降雨充沛的湿热带。它们分别是亚马孙河、刚果河和奥里诺科河,这些河流向海

洋排放的年平均水量超过1 000立方千米。

湿热带地区河流一年中的流量相对恒定,但在季节性气候占主导地位的地区,常年性河流的流量呈现明显的季节性。印度河的大部分水量来自喜马拉雅山脉,由于融雪的影响,夏季的最大流量是冬季最小流量的100多倍。对于主要流经区域处于高纬度和高海拔地区的河流,其最小流量常常为零,这些地区的气温一年中有一段时间低于冰点。在这些间歇性河流中,冬季冰冻期的最小流量与夏季融化期的大洪水形成明显对比,具有规律性和可预测性。

相比之下,常见于沙漠地区的暂时性河流,其流量具有间歇性和不可预测性。这是因为暂时性河流会对降雨做出响应,而众所周知,在许多沙漠中,降雨很难预测。一项对以色列内盖夫沙漠北部的河床研究表明,平均而言,该河道只在2%的时间内有水,也就是说,一年中大约7天有水。有些沙漠河流可能全年无流水。

沙漠气候中的河道流量的年际变化也是最大的,而湿热带常年性河流的流量的年际变化相对稳定。奎斯布河位于纳米比亚的纳米布沙漠中,该河中游的几十年流量资料显示,其每年有流量的天数介于0～102之间。

在更长的时段内,降雨和气温的变化也导致了河流水流情势的变化,尽管在许多情况下,人类干预使情况变得复杂(见第五章)。天然径流近期出现明显变化的一个示例发生在西非。撒哈拉沙漠以南的沙漠边缘带被称为萨赫勒,在20世纪最后几十年里,该地区经历了显著的气候干燥,并且这一趋势到21世纪仍在延续。巴克尔位于塞内加尔、毛里塔尼亚和马里边界交汇处附近,在该处进行的流量测验表明,20世纪末塞内加尔河的流量明显下降。1904—1992年,塞内加尔河巴克尔段的多年年均流量为716立方米每秒,而1972—1992年,其多年年均流量仅为379立方米每秒。1984年特别干旱,年平均流量为212立方米每秒。尼日尔河的情况和塞内加尔河类似。

有些河流很大,可以流经一个以上的气候区。例如,有些沙漠河流为常年河,因为它们的大部分流量来自沙漠以外降雨充沛的地区。这样的河流被称为"外源"河。尼罗河就是一个示例,澳大利亚的墨累河也是如此。由于蒸发和土壤下渗,这些河流在流经沙漠时水量损失很大,但因为水量巨大,它们依然可以保持连续流动直至进入大海。相比之下,许多沙漠外源河并不流入大海,而是流进内陆盆地。在南部非洲,来自安哥拉热带高原的水汇入奥卡万戈河,流入奥卡万戈三角洲。这个三角洲是一片巨大的湿地,位于博茨瓦纳北部的卡拉哈里沙漠。在中亚,来自帕米尔山脉的水经锡尔河和阿姆河这两条该地区主要的外源河流入咸海。

人们认为,有些河流非常古老。亚马孙河口附近沉积的泥沙表明,这条河流已经在南美洲流淌了1 100万年。在如此长的时段里,各种各样的因素当然都发生了变化,有些河流出现又消失了。海峡河的消失就是一个示例,大约20 000年前在欧洲西北部向西流淌的海峡河,如今沉没在分隔英国和法国的英吉利海峡之下。这是上一个冰期最寒冷时期的情形,那时世界各地的海平面比现在低得多,因为更多的水以冰的形式存在于水文循环中。当时,不列颠群岛的大部分以及整个斯堪的纳维亚半岛都被厚厚的冰层覆盖着,海峡河依靠英格兰南部包括泰晤士河和索伦特河在内的河流里的融水补给,这些河流已位于永久冰层之上。海峡河南面的其他支流包括塞纳河、索姆河、马斯河、莱茵河和易北河。

这样古老的河道并非仅具学术价值。南非的威特沃特斯兰德地区拥有世界上储量最大的金矿,这些金矿就是20多亿年前在水系中沉积贮藏的。当水流速度减缓时,河流挟带的金子就在砾石中沉积下来。地质学家称这些砾石为威特沃特斯兰德砾岩,目前已经从中开采出黄金近5万吨,占已开采黄金总量的40%,而未开采的部分,其可能储量依然占世界未开采黄金的三分之一以上。南部非洲的西海岸分布着宝贵的钻石矿床,而河流在这些矿床的形成过程中发挥了关键作用。在长达一亿年甚至更长的时间内,内陆矿床中侵蚀形成的钻石被瓦尔河和奥兰治河带到了海岸沿线。人们认为,这种河流搬运对沿海泥沙中钻石的质量也是有利的,因为这些宝石在搬运过程中往往会分解,高品质钻石的含量因而增加了。

侵蚀、搬运和沉积

水系塑造景观方式的一个重要衡量指标是"河网密度"。它等于河流干支流总长除以流域面积,反映了河道的疏密程度。因此,河网密度表达了河流切割景观的程度,能有效地反映地形特征。大量研究表明,不同地区的河网密度差异很大,取决于气候、植被,尤其是地质条件。植被稀疏的干旱地区、大雨频繁的温带和热带地区以及下方存在不透水岩石的地区,其河网密度值往往很高。

河流塑造地球大陆景观的主要方式有三种:侵蚀、搬运和泥

14

沙沉积。人们根据每一区域的主导性过程,利用这三种过程对单一河流和水系进行简单的三段式分类,即河源区、转移区和沉积区。

第一个区域由河流上游构成,大部分的水和泥沙都来自这一区域。河流侵蚀大部分都发生在这里,被侵蚀的物质经第二个区域输送,在第三个区域沉积。这三个区域是理想化的,因为每个区域都存在一定的泥沙侵蚀、储存和输送,但在每个区域内,只有一个过程占主导地位。

反映在图像测量中,河流上下游之间的坡度变化被称为"纵断面"。该断面从河源开始,到河口结束,由于河源部分地形陡峭,而在下游方向上坡度逐渐减小,因此纵断面通常呈凹形。这种通常呈平滑、上凹的形状有时会因出露的坚硬岩石而中断,从而形成局部陡坡。水流在这些区域形成急流,流速增大,侵蚀加剧,障碍物被长时间地消蚀。如果相对柔软的岩石上方覆盖着更加坚硬的岩石,那么这些地方可能形成瀑布。委内瑞拉的安赫尔瀑布是世界上最高的瀑布,当地人称之为克雷帕库派-梅鲁,它从异常坚硬的砂岩岩面上落下,高度达到令人惊叹的979米。

归根结底,河流挟带的所有泥沙都来自周围坡面的侵蚀和地表径流,但直接来源是河床和河岸。水流挟带的泥沙以三种方式存在:以溶液形式流动的溶解物,如钙、镁和其他矿物质;以悬浮形式存在的小颗粒;沿河床滚动、滑动或做"跃移"运动的较大的颗粒。当环境发生某种变化,如河床坡度减小时,此时河流能量以及挟沙能力降低,这些物质便开始沉降。其中的大部分物质沉积在大海中。据估算,在全球范围内,河流每年向海洋输送的悬移质约150亿吨,溶解物约40亿吨。

河流上游可能会流经基岩，但这种情况在下游不太可能发生。冲积河流的两侧是洪泛平原，河槽切入河流自身搬运并沉积下来的物质。洪泛平原是一片相对平坦的区域，在大流量期间会被周期性淹没，通常每一两年一次。当河水漫溢至洪泛平原时，流速降低，泥沙开始沉淀，在洪泛平原形成新的冲积层。

每片大陆上都可以看到特定型式的冲积河道，但归根结底都可以归为三种类型，即顺直型、蜿蜒型和辫状型。顺直型河道在自然界很少见，大多数情况下，它们是评估尺度的函数。在区域尺度上，它们可以被认为是顺直的，但在局部尺度上，它们都存在一定程度的弯曲或者蜿蜒。最常见的河道型式是被称为曲流（meanders）的一系列河弯，它得名于土耳其西南部以蜿蜒曲折而闻名的门德雷斯河（River Menderes）。曲流因凹岸侵蚀、凸岸沉积而发育。随着这些关联过程的持续发展，曲流可能会变得愈来愈弯，特别蜿蜒的曲流最终可能会在狭窄的颈部被切断，留下的旧河道变成了牛轭湖。冲积曲流在洪泛平原的纵、横两个方向上改道，其变化过程可以通过对比以往地图和重复拍照进行监测。横向改道是洪泛平原形成的一个重要过程。

众多水流分汊后再彼此汇聚而呈现辫状外观，这种河流被称为辫状河。这些多次交叉的水流中散布着许多小的且常常为临时的冲积岛。辫状河流通常分布在山区附近坡度相当陡峭的区域，一般挟带大量的泥沙。为什么有些河道蜿蜒曲折，而另一些呈现辫状？人们对此进行了大量的研究。影响河道型式的重要因素包括径流量和流速，而它们又与河道坡度和河道性质有关，特别是河床和河岸容易侵蚀，这会影响河流的泥沙供应。这些因素会随着时空的变化而变化。例如，北美的米尔克河在流

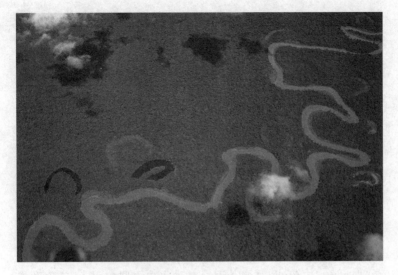

图4 新几内亚偏远地区的蜿蜒河流和牛轭湖

经加拿大南部的亚伯达省时是一条典型的蜿蜒河流,但在进入美国蒙大拿州后不久就突然变成了一条辫状河流。组成河床和河岸的物质不同,以及辫状河段河道拓宽降低了河流的动力,是导致这种变化可能的原因。

曲流裁弯形成牛轭湖是河道突然改变路线的一种方式,也是某些冲积河流的一个特征,通常被称为"改道"。这是一个自然过程,河流由原有河道改向,在邻近洪泛平原上形成新的永久河道,这种变化可能对人类活动产生重大威胁。在南亚印度河-恒河平原上,快速、频繁且常常重大的改道已经成为许多河流的典型特征。在印度,戈西河在过去200年里向西迁移了约100千米,甘达克河在过去5 000年里向东迁移了约80千米。巴基斯坦的印度河下游也曾发生过重大改道。河流突然改道的原因目前尚不完全清楚,但在印度河-恒河平原上,地震是原因之一。 18

改道有时会导致河道干涸，但在其他情况下，河道会分汊形成多河道河流。这些多河道河流被称为"网状"河或"分汊"河。乍看起来，网状河与辫状河很容易混淆，因为两者的型式大致相似。辫状河在单一河道中有多股水流，而网状河则有多个相互连通的河道。尽管如此，关于这些差异的争论仍在继续，同时河道型式的分类方法也是多种多样。河流流量的大小也会导致误解。在大流量情况下，有潜洲的辫状河看起来可能像单一河道；而在小流量时，网状河可能只在主河道中输水，所以看起来像单式河道。

大多数河流最终流入大海或湖泊，泥沙在那里沉积，形成三角洲地貌。三角洲（delta）这一名称来自希腊字母Δ，三角形或扇形是三角洲的典型形状之一。分别位于尼日尔河和尼罗河末端的非洲最大的两个三角洲就是这类三角洲的两个示例。

河流为三角洲的形成提供了泥沙，但影响三角洲形状的因素还有很多，包括径流量、输沙量以及流量、潮汐涨落和波能等相对重要的因素。尼日尔河三角洲和尼罗河三角洲这类扇形三角洲主要由海浪作用形成。河道水流主导作用下形成的三角洲通常延伸至更远的海中，形成三角洲瓣，其河道像鸟的脚趾或爪子一样分汊。密西西比河三角洲就是"鸟足状三角洲"的一个示例。潮汐作用主导形成的三角洲位于潮差较大或潮流较快的区域。它们的典型特征是许多岛屿被拉长并平行于主潮流，垂直于海岸线。新几内亚弗莱河三角洲和布拉马普特拉河与恒河交汇处的三角洲都是这类三角洲的很好示例。

河流末端沉积的物质可以在水下延伸至离开三角洲很远的地方，形成深海扇。世界上最大的深海扇位于远离恒河-布拉

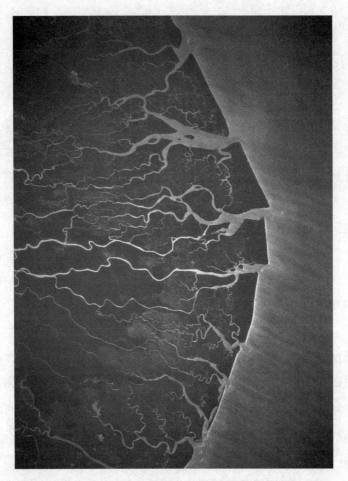

图5 尼日尔河三角洲是典型的扇形三角洲,也是非洲最大的三角洲

马普特拉河三角洲的水下。孟加拉深海扇长近3 000千米,宽1 000多千米,最深部分的厚度可能超过16千米。它通过海底峡谷与恒河-布拉马普特拉河三角洲相连,峡谷将河流中的泥沙输送到深海海床。孟加拉深海扇的起源可以追溯到4 000多万年前的一次构造事件,即印度次大陆与欧亚大陆发生碰撞,喜马拉

雅山脉在该事件中形成。

河流生态

　　各种各样的生物构成了河流生态,这张相互联系的生命网包括微小的藻类以及比人类还大的大鱼。形形色色的群落反映了各种不同的水流环境,从次大陆尺度流域中的大型平原河流到涓细、湍急的山间溪流。河流的物理结构对河流生态具有一系列影响,但河流的化学性质和生物性质也很重要,所有这些因素在一定程度上是相互关联的。水的含氧量、酸碱度、养分、金属以及其他成分主要由构成流域的土壤和岩石的类型决定,但在某种程度上也取决于它们与水中和陆地上的动植物的相互作用。

　　人们通常根据捕食和觅食方法来对河流生态系统中的生物进行分类。"撕食者"为取食小片落叶的生物,"刮食者"取食黏附在石头和大型植物表面上的藻类,"收集者"取食其他生物尸体分解后产生的有机物颗粒,"捕食者"吃其他生物。随着河流从河源向下游延伸,这些生物种群的相对重要性通常会发生变21化,反映了河道宽度、树荫遮蔽程度和水流速度等物理因素的变化。这就是"河流连续体理论",它描述了本质上以线性方式整合能源、食物网和河流级别的变化连续体。因此,小的河源溪流常常被上方悬垂的植被遮蔽,导致光照和光合作用受限,但植被的落叶贡献了有机物。在这些河段,撕食者和收集者通常占主导地位。然而在更远的下游,河流变宽,因而接受阳光变多,落叶变少,情况则完全不同。在这里,食物链通常以活的植物而非落叶为基础,所以撕食者很少,更多的可能是捕食者。

河流连续体理论是一个流行的模型,对许多河流生态系统的研究产生了影响,但它不是唯一的模型。在温带和热带地区,洪水将许多河流延展至其洪泛平原,研究河流生态的另一个重要模型因此强调每年洪水脉冲的重要性。"洪水脉冲理论"将关注的重点扩展至主河道之外,并更强调与洪泛平原常见的沼泽和湖泊等多种生境的相互作用。这些生境大体上与河流的河岸区(riparian,来自拉丁语*ripa*)是同义语,河岸区由经常影响水体或受水体影响的任一毗邻陆地组成。河岸区的植被在许多方面有助于维持水生生态系统的环境,其中包括维持岸坡稳定,并因此最大限度地减少侵蚀,过滤泥沙,处理流域内的养分,特别是氮。同时,河岸区树木上掉落的树枝或树干为许多鱼类和小型生物提供了木质生境区。

从生态角度看,河流的单向流动是一种独特的情况。流水影响着河流环境的许多方面,搬运其中的物质并因此有助于扩散生物和输送养分。流水影响着河槽的形状和河床的性质,湍急的水流会扰动河槽及河床,但也为河中的植物和动物提供了动态生境。同时,河流还向海洋生态系统输送水、能量、泥沙和有机物。绝大多数情况下这种流动是朝一个方向进行的,但也不是完全如此。例如,一些鱼逆流而上,迁移至上游产卵。从海洋迁移到淡水中繁殖的鱼类(即所谓的"溯河产卵"物种,如鲑鱼),就是最好的示例。鲑鱼主要在海洋中觅食生长,在河流中产卵后便会死亡,死后的残骸为水生生态系统和邻近的陆地生态系统提供了重要的养分和能量。

在河流生态研究中,重视流量的空间复杂性几乎是必然的,但流量随时间的变化同样重要。径流量、时间和流量变化创造了河流生物已经适应的五花八门的生境。例如,在地中海气候类型区域,河流的生态与流量的显著季节性变化相适应,因为大部分降雨发生在冬季(三个月的降雨量常常占全年的80%或更多)。凉爽多雨季节与炎热干燥季节的交替变换形成了河流洪涝与干旱的规律性变化,尽管季节旱涝强度的年际变化可能非常显著。

毫无疑问,河流生物在许多基本方面都依赖于河流的物理环境,尤其是气候、地质条件和地形。然而,这些关系也可以反向发挥作用。河流的生物组分也对物理环境产生影响,特别是在局部范围内。大型哺乳动物能够在许多方面彻底地改变河流的物理结构,河狸就是一个很好的示例。它们通过伐木造坝拦截泥沙和有机物,改变养分循环并最终影响许多其他动植物群落。

最后,值得再次强调的是,河流生态在许多方面的影响远远超出了河道本身。河流在塑造其流经的景观方面发挥了关键作用,水流也以同样的方式为在该区域栖息的许多动植物提供了重要服务,其中最明显的是提供了水和食物的来源。流水为生态系统输送和带走了许多重要养分和其他成分,但河流也产生了一些并非立竿见影的影响。许多陆生植物和动物物种的分布与主要水系的地理环境相协调,因为河流既可以作为物种传播的走廊,也可以作为生物传播的屏障。博物学家阿尔弗雷德·罗素·华莱士是最早认识到河流作为某些生物迁移屏障重要性的人士之一,19世纪中叶,他确定了南美洲亚马孙流域以主要河流为分界的明显不同的区域,每个区域中都存在独特的物种群落。河流作为屏障的观点是人们提出的众多假设之一,以解释亚马孙森林中惊人的丰富物种的进化起源。

亚马孙河：最壮阔的河流

亚马孙河几乎在所有方面都堪称最大的河流。其流域面积超过700万平方千米，是世界上最大的流域，占全球陆地面积的5%。在全球所有河流排入海洋的水量中，亚马孙河占近五分之一。亚马孙河流量巨大，甚至在远离河口125英里外的大西洋仍能发现亚马孙河的河水，早期的水手在远未见到南美洲大陆之前就能在海洋中喝到淡水。然而，亚马孙河下游的坡度非常平缓，以至于在其远离大西洋1 000多千米的上游仍能发现潮汐的实质性影响。

亚马孙河有约1 100条支流，其中7条支流长度均超过1 600千米。人们常常根据河水的颜色对这些主要支流进行分类，河水的颜色也反映了它们的来源。黑水支流的茶色源于低洼沙质土壤中析出的高浓度溶解植物质。白水支流的颜色源于来自安第斯山脉的大量泥沙。清水支流中挟带的少量泥沙和有机物来自圭亚那和巴西地盾的结晶岩。

在平原地区，亚马孙河干支流大多具有广阔的洪泛区，并伴有数千个浅水湖泊。整个亚马孙河流域多达四分之一的地区会被洪水周期性淹没，随着水位的上升，这些湖泊逐渐相互连接。根据GPS测量结果，研究人员发现，由于亚马孙地区洪水产生的额外重量，南美洲相当大一部分地区下沉了近8厘米，洪水退去后该区域又再次上升。这是我们观测到的最大的地壳年升降值。

亚马孙地区的许多动植物已经适应了季节性水涝的环境，有些地区的年水涝时间长达11个月，水深达13米。例如，亚马孙雨林的许多树种依赖洪水传播种子，这些种子或者漂向下游，或者通过以果实和种子为食的鱼类传播。亚马孙河水生生境的巨大多样性，在地球上最多样化的鱼类种群的产生过程中发挥了关键作用。科学家们已经登记了总共大约2 500种鱼类（可能还有超过1 000种尚待登记），亚马孙河的鱼类丰富程度稳超其他所有大型流域。其中两种最大的鱼，巴西骨舌鱼和丝条短平25 口鲇，每一种最大的都有大约200千克，是常人体重的两倍多。

按照长度这个衡量标准，亚马孙河一般被认为不是所有河流中最大的一个。在美洲，亚马孙河的长度稳居第一，但是在世界排行榜上，许多权威机构将尼罗河列为第一。然而，河流长度测量的难度意味着，关于这个问题的辩论无疑还将继续下去（见前文）。

玛瑙河：一条独特的河流

玛瑙河是南极洲最长的河流，只有32千米长，在许多方面都不同于世界上大部分地区的河流。玛瑙河位于麦克默多干谷地区，在被冰层覆盖的南极大陆沿岸，该区域是为数不多的几个无冰沙漠中的一个。这里的气候非常干燥且极其寒冷，年平均气温为−20℃。以干雪形式降落的少量降水（每年不到100毫米）对河流实际上没有直接影响，因为狂风使干雪无法落到地面。因此，玛瑙河和麦克默多干谷的其他河流每年有水的时间只有4～10周。这种情况发生在夏季，此时气温足够高，可以融化冰川冰，这是其河水的唯一水源。

玛瑙河自下赖特冰川流入万达湖，湖水的含盐度是海水的十倍以上，并覆盖着永久冰层。该地区没有植物，河流中没有鱼或昆虫,但河床上栖息着蠕虫、微生物和垫状藻类群落。这些藻垫可以在长期干燥的环境下存活,于是,玛瑙河在这一贫瘠的景观中成了相对的生命热点。

河流洪水

对水文工作者来说,"洪水"一词指的是河流每年的洪峰期,不管河水是否会淹没周围的景观[①]。然而,在更常见的说法中,洪水与河水漫过堤岸是同义语,这里用的就是这个意思。河流在正常情况下会泛滥,这种情况常常发生在洪泛区,正如洪泛区含义之所指,但洪水几乎可以影响整个河流。

极端天气,特别是强降雨或持续降雨,是引发洪水的最常见原因。冰雪融化是另一个常见原因。这些事件在一定程度上是可预测的,因为它们是季节性的。引发洪水的其他原因通常难以预见,它们包括山体滑坡、浮木堵塞、冰障、雪崩、火山爆发和地震。

河流洪水是影响人类社会最常见的自然灾害之一,经常造成社会混乱、财产损失和人员伤亡。洪水灾害促进了各种洪水预报技术的发展。洪水风险图通常用于土地利用分区,官方可据此禁止在极易发生洪水的土地上进行某些开发。预报洪水发生时间有几种不同的方法。大多数洪水的发生都存在季节性因素,常常可以使用气象资料进行预报,利用洪水流量过程线计算特定河流响应暴雨的洪峰滞时。

另外一种预报方法则是估算任何特定时段内平均只会超过

---

① "洪水"一词指的是江河水量迅速增减、水位急剧涨落的现象。——译注

一次的可能流量,因此使用了"50年一遇洪水"和"100年一遇洪水"等术语。一般来说,洪水的大小与其发生频率或概率成反比(换句话说,洪水越大,发生的可能性越小)。100年中可能仅发生一次的洪水,即100年一遇洪水,在任何一年中发生的可能性为1%,两次该量级的洪水之间的平均时间间隔为100年。就工程用途而言,了解特定量级洪水的发生概率是很有用的。例如,设计寿命为50年的桥梁要能抵御50年一遇的洪水,以防万一,常常要求能够抵御100年一遇的洪水。然而,这些都是统计学上的概率,大桥仍有可能被更大的洪水冲垮。

山体滑坡、冰川或岩浆流能够形成天然坝,堵塞河谷,这类成因造成的洪水难以预见。河流被天然坝堵塞后形成堰塞湖,会导致上游发生洪水,而天然坝溃决则会导致下游发生洪水。地震会导致巨大的山体滑坡,这是形成天然坝尤为常见的原因。例如,1968年5月,新西兰南岛的伊南加瓦地震引发了巨大的山体滑坡,堵塞了布勒河。上涨的河水向上回流了7千米,将河流水位抬升至高出正常水位30米。由于担心坝体可能发生灾难性的溃决,沿线所有居民全部撤离,但河水最终漫过了滑坡坝,坝体被逐渐侵蚀,下游并没有发生严重的洪水。

在过去的260万年中,也就是所谓的第四纪,大部分已知的最大洪水都是天然坝溃决造成的。据我们所知,最大的一次洪水发生在第四纪冰期,先前存在的大陆水系被冰盖堵塞,冰坝崩塌后形成巨大的洪水。在地球上发生过的巨大洪水中,有几次为发生在今天美国西北部的密苏拉洪水。18 000～13 000年前,堵塞今天的克拉克福克河的冰坝反复溃决,导致了大洪水。冰坝产生了一个巨大的湖泊,被称为密苏拉冰川湖,当冰坝周期性溃决时，湖水溢出，形成密苏拉洪水。据估算，密苏拉洪水的洪峰流量大到令人难以置信的程度，达 1 700 万立方米每秒，是当今世界上所有河流流量总和的十倍以上。

密苏拉洪水发生的证据是令人信服的，但它只是史前和地质时期已知发生或疑似发生的大洪水之一，有些洪水还未被充分证实。在全世界各种文化中，这类融合了事实和虚构的洪水传说成百上千。根据河道水流演绎而来的神话、宗教习俗和信仰为数众多，上面的故事只是其中的一部分，下一章我们将专门探讨这一主题。

29

神圣的水流

> 信道而且行善的人，他们的主将因他们的信仰而引导
> 他们；他们将安居于下临诸河的幸福园中。
>
> 《古兰经》第10章第9节

纵观历史，河道水流哺育了生命，使无数社会繁衍生息，但有时也会带来死亡和灾难。这是大自然的力量，既维系生命，也摧毁生命，这种双重功能在全世界不同群体中产生了文化共鸣。河流对人类的巨大影响力已经植根于历代无数的习俗、神话和宗教仪式之中。

神话中的河流

在希腊神话中，冥界或阴间被五条河流环绕。它们分别是阿刻戎河（愁苦河）、克塞特斯河（悲叹河）、弗莱格桑河（火焰河）、勒忒河（遗忘河）和斯堤克斯河（怨恨河）。人死之后，死者的灵魂进入阴间，在支付一定的费用之后，船夫将其摆渡过河

（有时是阿刻戎河，有时是斯堤克斯河）。在阴间，每一个初到者都要接受审判，经裁定善恶后被送入地狱或者等同于天堂的极乐世界。极乐世界的居民一旦忘记了前世，就有了重生的可能，而忘记前世则需要饮用勒忒河河水。

斯堤克斯河具有神奇的魔力。众神用其河水封印誓言，使之牢不可破。古希腊神话英雄阿喀琉斯年幼时被母亲倒提着浸入斯堤克斯河，除了脚后跟被抓住的地方外，整个身体刀枪不入。阿喀琉斯最终死于射中他脚后跟的一支毒箭，"阿喀琉斯之踵"一词就源于这一情节，至今还被人们用来形容一个人的致命弱点。

其他信仰体系中也存在渡过河流进入阴间的说法。在日本佛教习俗中，到阴间要渡过三途川，而在印度教的一些典籍中，则要渡过鞞多梨尼河（尽管只是针对有罪之人（那些在阳间做了善事之人不需要渡过该河）。

在希伯来人、基督教和伊斯兰教的传统中，河流在天堂的故事中也占有重要地位。早期基督徒接受了希伯来《圣经》及其中的《创世记》故事。在《创世记》中，从伊甸园中流出的一条没有名称的河流滋润着花园，并从那里分为四条河流。这四条流向世界不同地区的河流分别是底格里斯河、幼发拉底河、基训河和比逊河。虽然底格里斯河和幼发拉底河广为人知，但基训河和比逊河却使古代和中世纪的许多旅行者感到惊奇和困惑。比逊河曾长期与阿拉伯联系在一起，后来又被认为是恒河或印度河，有时也被认为是多瑙河。相反，基训河的源头通常被认为位于埃塞俄比亚，因此基训河被认为就是尼罗河。底格里斯河、幼发拉底河、恒河和尼罗河这四条相隔甚远的河流却在伊甸园

图6 16世纪出版的《教廷权威编年史》中天堂或者人间天堂里四条河
32 流的地图

30

中拥有共同的源头,对于这一显然不可能发生的事情,有人这样解释:最初离开伊甸园时,这些河流在地下流动,然后在离天堂很远的地方重新流回地面,而且彼此相隔很远。

《古兰经》中也详细描述了天堂里的四条河,天堂是安拉为虔诚的穆斯林准备的福地,常常被描述为"下临诸河的幸福园"。这四条河中,一条流淌的是水,另外三条分别流淌的是牛奶、葡萄酒和蜂蜜,供虔诚的穆斯林享用。人们常常认为,天堂里的四条河对伊斯兰园林的设计产生了巨大影响,这些园林是天堂在俗世的代表。许多伊斯兰园林按照四部分布局,由水渠分隔开来,渠水来自园林中心的水池或喷泉,但这种流水划分的四部分设计实际上早在伊斯兰教兴起之前就已经存在。因此,与其说它可能是反映穆斯林特有天堂观的布局,不如说它是对天堂的描述,反映了一种早就存在的园林形式。

在印度最早的一些梵文典籍中,河流占有重要地位。四大吠陀经典是印度教的根基,在第一部经典《梨俱吠陀》中,萨拉斯瓦蒂河是最著名河流中的一条,也被人格化为萨拉斯瓦蒂女神。在《梨俱吠陀》中,它是一条湍急的大河,但在《摩诃婆罗多》等后来的印度典籍中,它被缩小为一系列盐湖。在卫星影像的帮助下,一些对萨拉斯瓦蒂河感兴趣的当代学者近年发现印度塔尔沙漠中存在一些古老的干涸河道,并认为这些河道就是神秘的萨拉斯瓦蒂河。

洪水传说

在古今各种文化的神话中,大洪水故事出现的频率高得不可思议。《圣经·创世记》中描述的洪水在犹太–基督教世界广为人知,它与美索不达米亚地区早期巴比伦《吉尔伽美什史诗》中描述的洪水以及苏美尔和亚述时期的类似故事存在许多相似之处。洪水被解释为上帝清除地球上任性的人类的手段,虽然有一个人与其家人利用一艘船或方舟携带地球上的代表性生物成功地逃脱了洪水。在所有这些故事中,方舟最终停靠在山顶上,鸟儿被派去查看洪水是否退尽。大洪水具有重要的象征意义,既含有明显的净化元素,又是重生的载体,标志着前洪积世和后洪积世之间彻底脱离。在个人层面上,这一事件在各种圣水净化仪式中被有效地重复,包括基督教的圣洗圣事:在洗礼仪式中,受洗者在河水(或圣洗池)中洗去旧日罪孽并在基督里重生。这一仪式效法耶稣在约旦河中的洗礼。

在中美洲一些玛雅族群的书面记载中,也出现了类似的洪水导致的先前世界和新宇宙秩序的划分。在不同版本的记载中,洪水都发生在天上的凯门鳄被斩首之后,人们对此的解释是暴雨引发了洪水。在一些记载中,洪水中的少数幸存者使人类得以继续繁衍,但在中美洲许多其他洪水神话特别是阿兹特克人记载的洪水神话中,洪水没有幸存者,所以创世不得不重新开始。

北欧神话中的创世神话叙述了世界如何在火与冰的交汇处诞生,该交汇处是一个巨大的虚空,十一条河流流入其中。邪恶的冰霜巨人伊米尔诞生于此,并从他的左腋窝下诞生了第一个男人和女人。最终,众神杀死了伊米尔并用他的身体创造了世界。他的头骨变成了天空;他洒下的鲜血变成了挪威洪水,除了一个男人和他妻子乘坐一条用中空树干做成的船逃脱之外,其余冰霜巨人全部被淹死。

河
流

在澳大利亚众多土著群体的各种神话故事中,洪水也占据了重要地位,而沙漠景观中常常惊人的洪水至少可以在某种程度上解释这种突出地位。南澳大利亚维兰古人中流传着这样一个故事。故事的主人公是一个名叫琼班的唤雨巫师,有一天,在祈雨仪式中他有点心不在焉,结果唤来了一场异乎寻常的大雨。琼班试图警告他的族人,但是巨大的洪水来势凶猛,冲走了所有的人和财产,形成了一座淤泥山。这就是在山中发现金子和骨头的故事来源。

世界各地神话中描述的洪水可能基于真实事件,但这种可能性有时也引发了巨大的争论。例如,解构《圣经》洪水故事在19世纪地质学的兴起中发挥了核心作用。由于缺乏地质资料佐证,英国地质学家查尔斯·莱尔在其影响深远的著作《地质学原理》(共三卷,1830—1833年出版)中,摒弃了流行的诺亚洪水观点。在科学与信仰之争中,莱尔的书是我们解释世界起源主要根据的重要著作之一。它使人们普遍认识到,我们的星球远比神创论者认为的要古老得多。

圣 河

在许多信仰体系中,自然界的元素被赋予神圣的特征,其中一些特殊河流尤为突出。例如,对欧洲西北部的凯尔特人来说,河流是神圣的,许多河流被人格化为女神。今天在这一地区使用的一些河流名称可以追溯到居住在这些河流附近或死在其中的凯尔特神灵。在爱尔兰,博因河和香农河得名于从魔井中寻求智慧后淹死在河中的女神。

尼罗河对古埃及人的重要性体现在与河流相关的大大小

小的众神之中。女神伊西斯因丈夫被谋杀而在悲伤中流下的眼
35 泪化为每年的洪水，哈比神则是年年泛滥的洪水的化身。哈比
是尼罗河神，负责收集这些眼泪，住在今天阿斯旺附近的大瀑布
里，周围是鳄鱼和女神，其中有些是青蛙，其他的则是蛙头女体。
每年洪水开始泛滥时，埃及人都会为哈比献上大量的动物祭品。

　　在许多情况下，河流的神圣性与源于水这一太初元素的创
世神话有关。例如，人们认为加纳的比里姆河是阿基姆王国的
精神力量和源泉，因为传说阿基姆人诞生于该河的深处。事实
上，遍及非洲的河流、小溪和其他水体经常被视为神和祖先的
栖息地，因此倍受尊重。例如，在约鲁巴人的宇宙观中，最著名
的河神是尼日利亚奥贡河的统治者叶莫贾。叶莫贾是所有鱼的
母亲，也是孩子的赐予者，通常想成家的女性会用山药和鸡奉祭
她。在非洲南部的许多地方，居住在某些河潭中的精灵负责创
造传统治疗师（见下文）。

　　传统上，西伯利亚的许多土著居民也一直与自然有着密切
联系，河流和其他景观元素是其万物有灵信仰体系的核心。他
们认为，河流、泉水、湖泊和山脉皆有守护神，居民们必须定期通
过萨满祭司向这些守护神致谢并表达敬意。例如，在哈萨克斯
坦、中国和蒙古接壤的阿尔泰边疆区，卡通河被认为是阿尔泰土
著居民文化的核心。阿尔泰人认为卡通河是有生命的，并以不
同的方式向其表达应有的尊重，其中包括不要向河中扔石头，过
36 河时念诵一些特别的话语，以及晚上不要从卡通河中取水，因为
这可能会打扰河流精灵。

　　在西伯利亚西北部的秋明地区，曼西人对待河流的态度也
与其类似。亚尔比尼亚河等圣河禁止捕鱼，某些河段甚至禁止

河
流

划船，所以船只能从岸上拉着走。其他河段也有不同的禁忌：如禁取饮用水，某些河岸禁止伐木。河口被认为是亚尔比尼亚河最重要的部分，当地人经过时会向河里投钱。

在欧洲东南部，多瑙河水体在下游地区的保加利亚人和罗马尼亚人的传统丧葬习俗中扮演了重要角色。对于死亡是通往阴间的漫长旅程这一观念，多瑙河极具象征意义，并被融入往往很复杂的纪念仪式。为死者"放水"是为死者提供阴间用水的一种仪式。仪式细节在各个村庄有所差别，但其间通常都有一个孩子把河水送到特定的房屋。在保加利亚的莱斯科维奇村，送水的孩子是个女孩，送水之后，她和几名妇女一起回到多瑙河，在河岸上铺上桌布，摆上煮熟的小麦和红酒。妇女们点燃蜡烛，把给孩子的礼物挂在从苹果树上砍下的树枝上。女孩将右脚伸入河中并三次请求多瑙河见证这一仪式，然后将一个装有一支蜡烛、一些小麦和一片面包的空心南瓜从此处放入河中。当南瓜顺着多瑙河漂走时，死者就会获得阴间用水，但如果南瓜翻了，死者则会发怒。

河流是印度教最重要的圣地之一。大约 3 000 年前，在现在的印度地区，河流倍受吠陀时代雅利安人的崇敬。考古挖掘证据表明，选择良辰吉日在河流中集体沐浴的印度教习俗可以追溯到（并源于）吠陀时代之前 2 000 年的印度河流域哈拉帕文明的类似习俗。实际上，"Hindu"和"India"这两个词就源自印度河（Indus）。

几乎所有的印度河流都被奉为神灵，但印度河通常被称为印度七大圣河之一，其他圣河为恒河、亚穆纳河（或朱姆纳河）、萨拉斯瓦蒂河、戈达瓦里河、纳尔马达河和卡维里河。然而，印度河和卡维里河有时会被塔普提河和基斯特纳河取代。河流常常被认为是地球的血管,河道的许多特殊位置特别神圣,包括源头、河口和交汇处。在印度所有的圣河中,最神圣的是恒河。

恒 河

　　印度教徒与恒河的联系是说明河流神圣性的最鲜明示例之一。的确,在印度,"Ganga"既是恒河的名字,也是这条河化身的女神的名字。神圣的恒河已被庄严地载入许多印度史诗和经典,包括《罗摩衍那》、《摩诃婆罗多》、《吠陀经》和《往世书》。据说,圣人跋吉罗陀曾前往喜马拉雅山并最终成功地说服了恒河下落人间,完成了恒河下凡的壮举。在不同版本的故事中,控制恒河水流的都是湿婆神,湿婆的另一个名字叫作甘格特哈拉,或者"承接恒河者"。

　　对印度教徒来说,恒河水具有许多吉祥的特性。它能医治百病,在其中沐浴可以洗去虔诚者所有的罪孽。然而,最重要的是,当一个人的骨灰或骨头被投入恒河时,他的灵魂就会得到解脱而进入轮回。对许多印度教徒来说,圣城瓦拉纳西是实现这一最终转变的首选之地。恒河瓦拉纳西段西岸的迎水坡被划分为许多部分,每一部分都建有许多通往河水的长台阶,也就是"河坛",人们来这里沐浴、洗衣和火化逝者。每天大约有80具尸体在瓦拉纳西的两个主要河坛上火化,其中大部分尸体是从城外运来的。更多死者的骨灰被撒入恒河。一些尸体则不被火化,如曾经患有天花或死于霍乱的人,它们被缚上重物直接沉入圣河。玛尼卡尼卡是最重要的河坛之一,专门用于火化尸体,河岸上的井相传是毗湿奴在梵天创世时所挖掘的,毗湿奴是印度教三大主神之一，有时被描绘成人鱼。这里也是时间终结时所有造物或整个宇宙的焚烧之处。

恒河另外一个神圣的地点位于它与亚穆纳河的交汇处，一般称之为普拉亚格，这是一个朝觐圣地，靠近今天的阿拉哈巴德市。这里是印度教大壶节四大朝圣地之一。根据传说，这里也是神话中的萨拉斯瓦蒂河与恒河和亚穆纳河的交汇之处，因此更加神圣。大壶节每12年举行一次，数百万信徒在恒河中沐浴，清洗自己的罪孽。据悉，2001年的大壶节是有史以来世界上规模最大的盛会，参加人数大约6 000万。

神圣的河流生物

既然世界各地的人们对很多河流心怀崇敬，那么河流中的一些生物也成为尊敬和崇拜的对象就不足为奇了。在古埃及，尼罗河中某些种类的鱼是神话和迷信的一个主题，有一些种类在特定的居民区被视为圣物，而在其他居民区则不然。长颌鱼 具有易于识别的长且向下弯曲的吻部，奥克西林库斯城的居民视其为圣物，从来不吃它，该地区的墓地中曾发现大量的长颌鱼干尸。据希腊历史学家普鲁塔克所言，长颌鱼曾经引发了奥克西林库斯城居民和附近城市西诺波利斯居民之间的激烈冲突。原因是有一天，有人看到视狗为圣物的西诺波利斯人吃长颌鱼，于是奥克西林库斯城居民展开报复，他们捕杀了所有能找到的狗并吃了它们。

河豚在世界上的某些地区一直受到人们的尊敬。由于在《梨俱吠陀》中被提及，神圣恒河中的南亚河豚被赋予了宗教意义，成为历史上首批受保护的物种之一。在公元前3世纪，印度

图 7 大壶节，印度教徒大规模朝拜圣神圣的恒河

40

38

历史上最著名的统治者之一阿育王统治时期,它被赋予了特殊的地位。在东南亚,高棉人和老挝人都把伊洛瓦底江豚视为圣物,人们很少捕杀江豚。同样地,在南美印第安人的民间传说中,亚马孙河河豚也被认为是神圣的,人们相信捕杀河豚会带来厄运。在中国,长江白鱀豚被尊为"长江女神",直到2007年被宣布功能性灭绝。

不同种类的鲑鱼是重要的季节性食物来源,它们具有可以同时在海洋和淡水河中生存的能力,因而在一些社会的神话和仪式中倍受尊崇。大西洋鲑鱼在凯尔特人神话中具有特殊的地位。据说它和时间一样古老,知晓过去和未来的一切。爱尔兰传说《智慧鲑鱼》描述了传说中的猎人勇士芬恩·麦克库尔(英语化为费恩·麦库)早年生活中的一个重要片段。芬恩的导师是一名诗人,花了七年的时间寻找这条充满智慧的鲑鱼。当诗人最终捕获到智慧鲑鱼时,他让芬恩为他烹制,但芬恩烧鱼时被鱼烫到了大拇指,于是本能地把大拇指放进嘴里吮吸,从而获得了世界上所有的知识。

在中世纪的英国,人们认为鲑鱼是亚瑟王传说中无所不知的水中生物。格威尔是亚瑟王最优秀的部下之一,也是一名语言专家,在寻找猎人大师的过程中,他与许多聪明的动物交谈过。每一种动物都比前一种更聪明,其中最古老和最聪明的是塞文河上一个叫莱恩莱瓦的神话水池中的鲑鱼。据说,这种神奇的鲑鱼因为吞食落入池中的榛果获得了智慧。根据这个传说,鲑鱼背上斑点的数量代表了鲑鱼吞食的榛果的数量。

长期以来,鲑鱼一直是被欧洲殖民前的北美太平洋海岸等北纬地区土著群体的主要食物来源。作为主食,鲑鱼成了众多仪式、禁忌和神话故事的中心。随着鲑鱼大量拥入河流上游产卵，渔民们迎来了丰收季节，与只能靠干鱼干肉度日的季节形成了鲜明的对照，于是人们在这一季节伊始举行各种活动庆祝并表达对鲑鱼的崇敬之情。在20世纪的头几十年中，在该地区工作的人类学家曾详细记录了许多美洲土著群体举行的庆贺鲑鱼洄游季节开始的"第一条鲑鱼庆典"，后来这种仪式不再举行了。

在今天加拿大不列颠哥伦比亚省斯基纳河沿岸的钦西安人社区中，捕到第一条鲑鱼的渔民必须将四名负责的萨满请到现场。他们将鲑鱼放在雪松树皮做成的垫子上，然后在一名身着渔夫服装，右手拿着摇铃、左手拿着鹰尾的萨满带领下，将鱼送到酋长家里。在酋长家中，在社区年长成员的见证下，萨满绕着鲑鱼走四圈，然后身着渔夫衣服的萨满相继发出指令切下鱼头、鱼尾和清除内脏。仪式在吟诵荣誉称号的声音中举行，杀鱼使用的是贻贝壳做成的刀。人们认为使用石刀或金属刀会引起雷暴雨。

太平洋沿岸各地都有类似的第一条鲑鱼庆典，仪式大同小异，有些包括致辞和宴请，有些包括祭祀舞蹈。所有庆典都着重表达对鲑鱼的尊重以及对丰收的期许。捕捞季节人们食用新鲜的鲑鱼，它们还被晒干或做烟熏处理，作为冬天的储备食物。北美洲、西伯利亚和中国东北的许多族群也使用鲑鱼皮做衣服。

河流中的精灵

在世界各地的很多文化中，河流与各种神话生物之间存在关联是很常见的。在德国和北欧民间传说中，这类水精灵被称为"nixie"（单数nix），通常怀有歹意。它们经常引诱人类受害者接近，将他们诱入无法逃脱的水中。水精灵的化身可能是男性，也可能是女性。德国的民间传说中最著名的水精灵之一是罗雷莱，她是一位美丽的仙女，坐在莱茵河的一块岩石上（莱茵河也是以她的名字命名的），用动人的声音或梳头的优雅动作常常将渔民诱入险境。斯堪的纳维亚民间传说中的水精灵更多是男性，他们用小提琴演奏迷人的曲子，把女性受害者引诱到河里或湖中。孕妇和未受洗的儿童尤其容易受到伤害。

在斯堪的纳维亚民间传说中，另一种河流精灵叫"bäckahästen"或"溪马"，这是一种威严的白色野兽，经常出现在河岸上，特别是在雾天，引诱疲惫的旅行者骑乘。人只要骑上马背就再也无法下马，然后溪马就会跳进河里淹死骑马的人。苏格兰民间传说中的马形水鬼是溪马的直接对应物。马形水鬼最常见的形象是一匹俊美且温顺的马，但也可能是一个毛发旺盛的男人，手像钳子一样可怕。他会躲在河岸上，当不幸的行人路过时便跳出来把他撞死。

在日本，与河流有关的精灵叫河童，这是一种常常被描述为介于孩子和猴子之间的生物，喜欢恶作剧。河童最喜欢的把戏之一就是把人、马或牛诱入河中淹死。对河童及其习性的描述存在区域性差异，但它最常见的特征之一是喜好黄瓜（通常被认为是繁殖能力的象征）。在日本的某些地区，人们相信游泳前吃黄瓜肯定会受到河童的攻击，但在其他地区这却是一种确保免受河童攻击的方法。不管怎样，与河童有关的许多节日中都有黄瓜祭品，而在现代日本，河童和黄瓜之间的联系通过"河童细卷"（一种黄瓜做的寿司）的名称延续了下来。有趣的是，在过去百年左右的时间里，经过人们的改造，有几百年历史的河童从邪恶且令人讨厌的水神变成了一种无害而且可爱的吉祥物。作为一种全国公认的象征，河童被用于为唤起人们对日本乡村历史的怀念而举办的各种活动。具有讽刺意味的是，其中的一项活动是以城市河流周围环境再生为宗旨的清洁水运动，该活动呼吁清理河流，以便河童回归。

传统上，河童与水精灵、溪马和马形水鬼一样，都是邪恶的引诱粗心者入水并致其死亡的河流精灵。相比之下，在非洲南部科伊桑语和班图语土著民族的传统宇宙观中，与水系和其他水体有关的精灵的行为截然不同。许多群体将喜欢生活在水中特定地点的精灵视为祖先。在河流中，这些地点通常为瀑布下方的深潭，"有生命"且快速流动的潭水经常会产生很多泡沫。这些水精灵化身为各种动物，主要是蛇和美人鱼。它们在许多方面与人类互动，其中最重要的一个方面是它们对传统疗法及其从业者至关重要。

传统上，水精灵召唤某些被选中的个体成为占卜者或治疗师，这一过程通常包括将入选者的身体在某个水塘中浸没几个小时、几天甚至几年。当身上缠着蛇的男子或女子从河流深处出现时，他或她已经获得了超自然的能力和治疗技能，包括药用植物的知识。这种被带到水下的经历可以发生在梦中，但这仅仅是一种通知——祖先正在召唤此人成为治疗师。这种召唤通常发生在生病一段时间后，而孩子则往往是在河边玩耍时被召唤。抵抗这种召唤是不明智的，通常会招致不幸。对落水失踪或者可能永远也回不来的被召唤者，亲属不能表现出悲伤情绪。

在南部非洲的土著社区中，水精灵具有崇高的地位，人们对许多河流、水塘和水源地既敬畏又恐惧，还心怀尊敬。这种神圣性导致在使用和进入这些区域时存在许多禁忌。通常情况下，只有治疗师、国王和酋长可以进入这些区域。由于担心被带入水中再也无法回来，普通民众被禁止接近神圣的水池。这些禁 ⁴⁵ 忌只是河流对人类错综复杂的影响中微不足道的一部分，而这种影响力可以追溯到人类的起源。河流的流动在诸多方面塑造了历史，下一章我们将详细探讨。

46

流动的历史

> 密西西比河是泥水，圣路易河是清水，而泰晤士河是流
> 动的历史。

约翰·伯恩斯（1858—1943）

（英国政治家）

河流既反映历史，也有助于创造历史。社会与河流相互作用的原因很多，这些动机可简单地分为两类：一类基于河流的有益方面，一类反映河流的危害。人类从河流中获益颇多。我们从河流中捕鱼并食鱼的历史已有数万年。河流为我们提供了生活、工业和农业用水，还为我们提供了各种矿物，包括黄金、钻石以及沙子和砾石等重要的现代建筑材料。人们可以利用河流蕴含的能量发展贸易和旅游、发电以及清除人类活动所产生的各类废弃物。河流是人们进行休闲和审美活动的理想去处，也是野生动物的庇护之所。相反，河流也会带来惶恐和惊惧。这与水量有关：太多则发生洪涝，太少则缺水。河流水质也会引

起人们的担忧,因为河水能传播疾病或引起砷等矿物质的危险富集。

　　人类社会所感知的河流的所有这些方面都不可避免地对人类历史进程产生了一定的影响。欧洲的历史在很大程度上就是多瑙河的历史。孟加拉国众多的河流既构成了国家的景观,也与人民的生活息息相关。没有泰晤士河,就没有伦敦。河流是许多社会及其历史的重要组成部分。

最初的文明

　　3 500～5 500年前,在世界多个地区的大河的洪泛平原上诞生了古代文明。底格里斯-幼发拉底河、尼罗河和印度河的冲积河谷沿线分别出现的苏美尔文明、埃及文明和哈拉帕文明,很大程度上源于这些河流提供的关键的益处:充足的淡水供应、肥沃的冲积土壤以及现成的贸易和旅行运输通道。在上述三个示例中,干旱的地理位置强化了社会对河流的依赖,使得农业生产以及维持沙漠环境下的持续生存特别依赖可靠的水流。这三个水系都是外源性的,即水系发源于气候湿润的地区,可以维持水流常年流经沙漠。

　　一种将产生这些最初文明所涉及的诸多因素联系在一起的理论认为,随着大量人口聚集在同一个地方生活,管理沙漠地区灌溉所需的中央集权组织也使得发展复杂社会成为可能。这种趋势最终导致了早期城市的形成,以及人们普遍认为的最初的文明。这三个起源于河流的早期文明都发展出了各自的引水、导流、种植和储存食物的方式。这三个地区还分别出现了文字体系、立法体系和其他许多文明特征。这一"大河文明"理论表明，这些早期社会对河流的人工控制和调节是文明内在和必要的先决条件。

另一种观点则进一步深化了这些早期复杂社会与河流之间的联系，认为文明的性质、特征和寿命在一定程度上是河流性质的反映。底格里斯-幼发拉底河、尼罗河和印度河都是大型外源性水系，但在其他方面却截然不同。尼罗河在埃及境内的坡度相对平缓，在过去的几千年中，河道只发生了曲流裁弯和小幅东移等不大的变化。尼罗河的洪水泛滥通常是定期且可预测的。埃及文明的稳定和长期延续可能是其河流相对稳定的反映。相反，印度河陡峭的河槽经历了印度河下游平原上数次远距离重大改道，以及喜马拉雅山脉冰川坝溃决所引发的数次特大洪水。在对于包括摩亨佐·达罗在内的许多哈拉帕城市被废弃的各种可能解释中，学者们考虑到了大洪水和/或河流改道导致供水中断造成破坏这一可能。

位于美索不达米亚冲积平原上的苏美尔文明曾经面临同样的河流改道的问题。美索不达米亚意为"两条河流之间的土地"，因其众多城邦的兴衰而闻名。这些城市大多位于幼发拉底河沿岸，可能是因为幼发拉底河在灌溉方面比流速快、水量大的底格里斯河更容易控制。然而，幼发拉底河是一条多河道分汊再交汇的网状河流。随着时间的推移，在其他河汊形成后，个别河汊不再流水，这些河汊沿线上的定居点不可避免地减少并因供水枯竭而被废弃，而其他河汊沿线的定居点则因水量充足而扩张。

探索的通道

人类一直在使用河流提供的直接通道探索新的土地。考古

48

河流

46

证据表明，在旧石器时代或石器时代早期，早期人类沿着岛屿上的主要河流进入后来人们所称的不列颠岛，之后逐步扩散并在离河岸较远的地方定居。同样，大约6 000年前，新石器时代的部落沿着河道从东南地区进入中欧。在这两种情况下，河谷为这些早期定居者提供了丰富的基本资源：水、鱼和猎物丰富的洪泛平原。

数千年后，北美洲的巨大水系使欧洲先驱们得以在新大陆上开疆拓土，开辟贸易并最终进行殖民统治。16世纪，在同胞雅克·卡地亚1530年代进行圣劳伦斯河探险之后，法国商人、探险家和传教士成为首批陆续到达五大湖区的欧洲人。他们受法国国王派遣，主要任务是绘制水系图，这些水系是通往新大陆的快速通道。这些河流通常是穿过北美洲原本无法穿越的森林的唯一通道，通航期可使用独木舟，冬季许多支流结冰时则使用雪橇。

1804年，当美国总统托马斯·杰斐逊派遣梅里韦瑟·刘易斯和威廉·克拉克去考察、勘测刚从法国人那里购买来的北美洲大片土地（路易斯安那购地案）并形成报告时，河流仍然是最容易走的路线。刘易斯和克拉克带领探险队沿密苏里河溯河而上，越过落基山脉，沿哥伦比亚河顺流而下，最后到达太平洋。他们的探险和带回的资料，特别是关于太平洋西北地区的资料，在美国向西扩张的过程中发挥了关键作用。

在俄罗斯扩张其势力、扩大对西伯利亚影响力以及西欧列强渗透非洲的过程中，沿河输送也发挥了类似作用。河流作为殖民探索通道的重要性不仅仅是历史关注的话题。在18世纪中期美洲殖民化以及19世纪后期帝国主义向非洲和亚洲扩张期间，河流通常被用作边界，因为它们是欧洲探险者在地图上绘制的第一个通常也是唯一的特征。那些与竞争对手谈判殖民地领土分配的欧洲外交官对所要瓜分的地方知之甚少。通常情况下，他们有限的了解仅源于地图，而这些地图很少显示细节，河流是标明的唯一明显物理特征。今天，许多国际河流边界就是这些基于贫乏地理知识的历史决定的遗留物，因为各国一直都不愿意改变原始划界协定中的领土边界。

澳大利亚的墨累河

墨累河位于澳大利亚东南部，它对于澳洲大陆的文化、经济和环境都具有重要意义。如果算上它两条最大的支流马兰比吉河和达令河，其重要性就更大了。墨累-达令河流域的面积占澳大利亚国土面积的14%。在欧洲人到来之前的数千年里，许多土著居民一直依靠这条河流提供的丰富资源生存，他们利用长在河岸边的桉树的树皮制成独木舟，沿河进行狩猎并开展贸易。岩画、考古遗址和墓葬遗址都是反映这些早期居民生活的证据。他们从河流中获取各种食物，包括鱼、淡水龙虾、贻贝、青蛙、海龟、水鸟和鸟蛋。

直到1820年代，欧洲探险家才第一次见到墨累河。查尔斯·斯特尔特船长沿着马兰比吉河进入墨累河，发现了墨累河与达令河的交汇处，然后继续向下游航行到达墨累河入海口。南澳大利亚殖民地的建立与斯特尔特的探险故事在伦敦的出版有一定关系。早期的欧洲定居者开始沿墨累河向澳洲大陆内陆渗透，河岸沿线开始出现小型定居点和牧羊场。墨累河在欧洲历史上令人印象最深刻的标志之一是明轮船，众多的明轮船在墨累河水系上来回穿梭，运输羊毛、小麦和其他货物，促进了墨

51

累-达令河流域的开发。始于1887年的灌溉农业加速了定居点的建设和河流供水系统的开发。

如今，墨累-达令河流域是澳大利亚最重要的农业产区，粮食产量占全国总产量的三分之一以上。它拥有全国65%的灌溉农田，养活了全国四分之一以上的牛群和近一半的羊群。它还向堪培拉和阿德莱德等主要城市供水。然而，天然状态下的墨累河水量变化极大且极难预测。在严重干旱期间，它不再是一条河流，而变成了一连串的咸水坑。多年以来，人们一直在对其进行流量人工调节，以确保供水。随着一系列水利工程和技术措施的实施，墨累河得到成功治理。这些工程包括五座大型水库，即达特茅斯和休谟两座大坝，以及马尔瓦拉、维多利亚和梅宁德等人工控制湖泊。自1936年休谟大坝竣工后，整条河流一直保持着连续流动。通过修建一系列隧道和管道，墨累河和马兰比吉河还从雪河获得一部分水量。为了进一步提高调水能力，建设了由13座堰、闸构成的流量调节系统，并在河口附近建造了5座堰坝，以防止海水倒灌。墨累-达令河流域存在大量天然状态的盐，生活和农业用水的水质一直是个问题。因此，相关部门制定了一系列盐分截留方案，防止盐分进入河流。这些方案包括建设拦截盐水并进行蒸发处理的大规模地下水抽水工程 52
和排水工程。

天然的屏障

许多例子表明，河流是群体间互动的天然屏障，在某些情况下，群体间持续分离的时间相当长，以至于出现了遗传学意义上的显著差异。在中非的灵长类动物中，倭黑猩猩和普通黑猩猩之间存在一条明显的分界线——刚果河。倭黑猩猩只生活在刚果河以南，而普通黑猩猩只生活在刚果河以北。黑猩猩不会游泳，因此刚果河有效地分隔了这两个群体，基因分析表明，这两个群体拥有共同的祖先，分隔时间约130万年。

河流对某些人类群体之间基因、文化和语言的流动所具有的类似屏障作用已经得到证实。在新几内亚高地，拉马里河似一道鸿沟标示出两岸的福尔人和安加人在文化与语言方面存在的巨大差异，人类学家将此作为典型案例记录了下来。这两个群体说着完全不相关的语言，拥有明显不同的文化，也是不共戴天的敌人。尽管存在这些差异的原因可能有多种，但拉马里河及其陡峭河谷所形成的难以逾越的天然屏障无疑是一个重要原因，尤其考虑到福尔人认为人是不会游泳的。

河流总是划分出这样的边界，这些边界既有真实的，也有想象的。公元1世纪，罗马帝国将欧洲的多瑙河作为帝国北部边界，因为它容易防御，因此也在"文明"帝国和对岸野蛮部落之间划出了一条清晰的界线。在今天的欧洲，多瑙河是斯洛伐克和匈牙利之间的国际边界，同时也是罗马尼亚与塞尔维亚、保加利亚和乌克兰之间的边界。在德国南方很多巴伐利亚人的心目中，多瑙河是南北方不同文化之间的象征性边界，他们亲昵地称其为"韦斯伍斯特赤道"（字面意思为"白肠赤道"，白肠是德国南部最受欢迎的一种食品）。

世界上不下四分之三的国际边界是河流，至少也是部分河段，这反映出了河流作为天然屏障的重要性。然而，河流边界不稳定也是众所周知的，移动是河流的固有倾向，这种倾向会给河流两岸的对立国带来法律、技术和管理层面的各种挑战。

53

这些挑战包括在一个不断变化的自然要素中确定一条明确的界线,以及管理跨境水资源分配。这两个因素当然是相关联的:确定边界的确切位置影响到对水域本身以及如何利用(例如航行)或滥用(例如污染)水域的法律权利。通常情况下,确定河流边界遵循"深泓线"(河流中最深的河槽)原则,但也使用其他原则。有些边界是河道两岸之间的中间线或转折点之间的连线,另外一些则是以河岸一侧为边界。有时,基于机会主义的考量,两个国家可能会选择不同的法律原则来确定边界位置,由此产生的争端往往需要通过国际仲裁来解决。即使两国就边界问题达成一致,侵蚀和沉积作用还是会改变河岸、中间线或深泓线,从而使一国受益而另一国受损。

墨西哥与美国成立的双边机构"国际边界和水务委员会"就是诠释河流边界认定与河流管理之间关系的一个历史悠久的案例。该委员会成立于1884年,它本来的职责是划定两国陆地边界并确定蒂华纳河、科罗拉多河和里奥格兰德河的边界位置,但是1944年,该委员会又被赋予分配里奥格兰德河水资源之职责。今天,国际边界和水务委员会把大部分时间花在水管理和水配置上,而不是边界的确定上。然而,并非所有的河流边界争端都能通过和平谈判方式解决。1969年,苏联和中国因乌苏里江国际边界问题,特别是珍宝岛主权问题发生了一场持续数月的激烈冲突。

54

河权与冲突

淡水作为一种重要的资源,加上其在河流、湖泊和地下含水层中的地理分布不均匀,不可避免地导致了不同群体在水权问题上的政治纷争。有时，在共享水资源的权利上的分歧会导致军事对抗，所谓"水战争"可能成为21世纪冲突主要根源的观念已在某些学术圈、媒体圈以及政治言论中变得相当普遍。

许多河流跨越（以及沿着）国家之间的边界流动，世界上大约60%的淡水来自多个国家共有的河流。一些较大的流域为多个国家所共享。多瑙河在这方面可谓首屈一指，其流域为至少19个欧洲国家所共享。另外5个流域——刚果河流域、尼日尔河流域、尼罗河流域、莱茵河流域和赞比西河流域——由9～11个国家共享。这些事实表明了可能的河权问题的规模，虽然存在多个利益相关者绝不是产生政治不满的必要条件。例如，恒河只流经两个国家，但在1975年印度建成法拉卡堰后，依然导致了印度和孟加拉国长达20年的对峙。孟加拉国抱怨说，由于距边境上游约18千米处的法拉卡堰的分流，他们被剥夺了灌溉用水，并且面临着日益严重的盐碱化问题。

最终两国于1996年签署了一项水资源共享协议，但是类似恒河这样的争端并不罕见。反对上游国家排放污水、修建水坝或过度灌溉，是跨境河流问题上存在分歧的典型理由，因为这些行为会导致下游国家可用水量减少或水质降低。其中许多争端通过国际条约得到了和平解决，但还有许多争端没有解决。此外，并非所有国际条约都旨在处理国际河流争端中所有相关方的诉求。这里以尼罗河为例。埃及和苏丹达成了一项国际协议，规定了允许通过阿斯旺大坝的尼罗河水量，但尼罗河流域其他八个国家均未就尼罗河水的使用达成协议。鉴于埃及和苏丹是尼罗河流入地中海前流经的最后两个国家，与上游国家就水权问题达成类似协议似乎是可取的。但到目前为止，敲定这样一项条约的细节已被证明是一个无法完成的任务，尼罗河水权问题上的分歧依旧是引发该地区许多政治问题的根源。

通过协议解决水资源争端的做法由来已久。国际水法的起源至少可以追溯到公元前2500年，当时，拉加什和乌玛这两个苏美尔城邦达成一项协议，解决了中东底格里斯河上一条支流的水资源争端。由于在当代缺乏协议，水纠纷仍然是底格里斯-幼发拉底河流域一个重要的潜在冲突根源。虽然该地区目前水资源过剩，但规划开发的规模还是引起了人们的担忧。土耳其的东南安纳托利亚工程是一项针对两河河源地区的区域性开发规划，计划最终建造22座水坝。1990年，当阿塔图尔克大坝拦截幼发拉底河水流、坝后的水库开始蓄水时，叙利亚和伊拉克立即对此表示担忧，尽管两国政府事先都得到了警示，且截流之前土耳其已经加大下泄流量进行了水量补偿。预计到2030年左右，东南安纳托利亚工程将全面完工，届时幼发拉底河的流量将减少60%，严重危及下游叙利亚和伊拉克的农业。底格里斯河和幼发拉底河沿岸三国一直试图就这两条河流的用水问题达成协议，而目前对这一协议的需求变得越来越迫切。

中东其他地区已经因水资源短缺发生了武装冲突。1950年代和1960年代，以色列、叙利亚和约旦之间因试图从约旦河和雅穆克河引水问题发生了多次军事冲突。1967年，就在以色列与其阿拉伯邻国之间发生"六日战争"之前，时任总理列维·埃什科尔宣布，"水是一个事关以色列生存的问题"，以色列将动用"一切必要手段确保河水继续流动"。

从那时起，在人们对共享流域国家特别是中东国家之间的关系未来如何发展的普遍看法上，"水战争"的幽灵变得愈加显现。然而，并非所有政府都认为更多的国家间冲突是不可避免的，他们甚至也不认为这种冲突是跨境河流管理中最重要的方面。国民经济发展只是"水安全"的一个方面。"水安全"是指能够可持续地获得数量足够、质量可接受的多用途用水，这些用途既包括社会和文化需求，也包括维持生态系统功能这一重要方面。所有这些用户都应该拥有对河流的权利，因此用户之间的冲突可能出现在主权国家以外的其他层面上。

湄公河流域的情形就说明了这一点。与世界上许多国际流域一样，它既被视为区域经济发展的引擎、民生资源的重要基础，也被视为保护生物多样性的重要场所。1995年，湄公河流域的四个国家达成了一项协议，批评人士认为该协议过分强调了湄公河的巨大水电潜力以及灌溉储水能力。这种潜力的开发不可避免地聚焦于国家层面，往往会得到银行和其他政府等国际发展伙伴的支持。有人指出，这些国家和组织认为湄公河的资源没有得到充分利用且适合开发，而这种立场忽视了依靠湄公河维持生计的当地资源用户的活动，这是令人担忧的。

泥土中的历史

河流中多年沉积下来的泥沙记录了流域各个时代发生的变化。泥沙分析是自然地理工作者解读景观历史演变的一种方法。他们可以研究泥沙本身和/或其包含的花粉或孢子等生物遗骸的物理和化学特征。由于悬崖等自然因素或者人类的作用，泥沙可能暴露在自由面上，在这些地方进行检查和取样非常方便，但大多数情况下，使用样芯钻取装置取样时，顺序是从上到下、逆时间的。

河流中物质沉积的简单速率可以很好地反映流域条件的变化。例如，对流入北美洲东海岸切萨皮克湾的布斯河300年来的泥沙沉积研究表明，由于该地区土地利用的变化，流域土壤侵蚀量发生了显著变化。人们认为，在欧洲人于17世纪中叶开始在布斯河流域定居之前，土著居民对流域没有产生显著的环境影响，1750年之前的泥沙沉积率约为每年一毫米。但是，随着第一批欧洲人开始早期的森林砍伐和农业种植，到1820年，泥沙沉积率已增长了八倍。在接下来的100年里，随着树木砍伐量的增加和农业的发展，土壤侵蚀随之加快，沉积率于1850年达到峰值，约每年35毫米。在更近的时期，自1920年开始，由于城市化和修建水坝的共同影响（前者保护了土壤，后者阻止了泥沙的输送），土壤侵蚀和泥沙淤积减少了一个数量级。泥沙沉积率几乎降到与前欧洲殖民时代普遍存在的本底条件持平的水平。

河流泥沙中经常发现大量来自周围植物的花粉，花粉分析可以提供很多关于以往区域条件的信息。植被类型可以被各种因素改变，这些因素既可以是布斯河流域示例中的人类干扰，也可以是气候变化或土壤条件变化等完全自然的原因。利用从湖泊和沼泽中提取的极长泥沙样芯，我们可以重构相当长时期内植被的变化，在某些情况下甚至超过100万年。由于气候是植被的重要决定因素，花粉分析也被证明是追踪过去气候变化的重要方法。

例如，一项对取自日本琵琶湖底的250米长样芯的重要研究表明，在过去的43万年中，花粉发生了变化，在这一时期可以识别出5个冰期—间冰期周期。在冰期，松树、桦树和栎（或白栎）树的花粉占主导地位，表明当时的气候凉爽温和，倾向于亚

58

第三章 流动的历史

59

北极气候。间冰期则相反,包括落叶紫薇等阔叶树以及常绿锥(一种水青冈)在内的暖温带气候典型物种的花粉值较高。

在更大尺度的景观元素中可以发现环境变化的其他证据。例如,许多现代河流的洪泛平原上都留有以前河道的痕迹,也就是所谓的"古河道",它们在规模和/或形态上都与现在的河流不同。如果古河道被埋藏在较新的泥沙之下,那么它可能是由流向较低基准面的河流形成的,这表明当地的海平面或湖泊水位随后发生了变化。人们认为,存在于许多河谷中的河成阶地反映了气候的波动,气温和降水对河流活动的直接影响和间接影响推动了这些阶地的形成。

水　力

两千多年来,人们一直利用流水和跌水的能量驱动水轮做功。流水驱动巨大的轮子,连接在轮轴上的传动轴通过由齿轮和嵌轮构成的系统,将水的动力传递给玉米磨盘等作业机械。较早描述碾谷水磨的人是罗马工程师维特鲁威,他编写的十卷本著作涵盖了罗马工程的各个方面,东地中海地区与这项技术的首先使用密切相关,尽管同一时期中国也出现了自己的水力利用传统。罗马的水轮经常与其他形式的水利工程配套使用,如输水渠和水坝等,用来输水及控制流经水轮的流量。今日以60色列的凯撒利亚·马里蒂玛镇位于克洛科迪翁河附近,切姆彤市和泰斯图尔市位于北非罗马玉米带(今属突尼斯)的迈杰尔达河(古代巴格拉达斯河)沿岸,这几个地方目前还有多套罗马水磨在进行大规模磨粉加工。紧靠凯撒利亚·马里蒂玛镇外的水磨坊由四个垂直的水轮组成,由水坝的输水渠供水。

在古代，河流动力被广泛用于碾磨谷物，但也具有其他用途。水磨还被开发用来驱动杵锤和锯子，用于捣碎矿石和切割岩石。在中世纪的欧洲，各种水力机械越来越普遍，体力劳动被逐渐取代。中世纪早期的水磨可以做 30～60 人的工作，在 10 世纪末的欧洲，水轮被广泛应用于各个行业，包括驱动锻锤、榨油机和织丝机、甘蔗破碎机、矿石破碎机，在制革厂破碎树皮，敲打皮革以及碾碎石块。尽管如此，大多数水磨仍被用于碾磨谷物，以制作各种食品和饮料。《末日审判书》是英国 1086 年编制的一份调查报告，其中载明水磨坊 6 082 家，不过这可能只是保守估计，因为遥远北部的许多磨坊并没有被记录在案。到了1300 年，这一数字已上升到 10 000 以上。

在整个欧洲，水磨坊一般属于领主、城市企业、教堂或修道院。西多会修道院在 12 世纪晚期英格兰"缩绒"水磨坊的最初发展中发挥了重要作用。缩绒或毡化是修道院庄园生产羊毛布料过程中的一道工序。它包括织物纤维的洗涤和固结，这两者都是羊毛布料恰当修整不可缺少的步骤。水力技术的引入革新了缩绒工艺，在此之前该工艺一直依靠人力击打布料。例如，在英格兰南部的怀特岛上，第一座缩绒磨坊就建在夸尔修道院的西多会修道院中，它坐落在修道院庄园大片牧场附近的一条小溪上。从羊群上剪下的羊毛在修道院加工成布料，然后卖到附近的城镇。

中世纪的水磨坊通常使用水坝或堰集中跌水并堵水成池来驱动水轮。对河流的这种改造在整个欧洲变得越来越普遍，到中世纪末期，也就是 15 世纪中叶，相当多的河流和小溪旁都建起了水磨坊。工业革命时期一系列的发明改变了英国的棉花加工方式并催生了工厂体制这一新的生产方式,而在此之前水力一直占据着重要地位。早期的纺织厂使用水轮驱动机器来生产布料,所以它们常常被称为织坊(mills)。

尽管河流继续在工业发电中发挥作用,但水力的优势地位很快就被燃烧木炭、煤以及后来的石油和天然气产生的蒸汽动

河
流

图8　大约1850年,英格兰北部曼彻斯特艾威尔河上的棉纺厂,这里的河流和运河中的水流是工业革命的重要组成部分

力所取代。所有的热电站,无论其使用的热源是化石燃料、核能还是地热能,都是将水或其他流体转化为蒸汽来驱动涡轮发电机。蒸汽必须在冷却系统中冷凝才能在涡轮机中再循环,因此需要大量的冷却用水。大部分冷却水来源于河流、湖泊、地下含水层和海洋。

在现代,随着水力发电技术的出现,河流流水的潜能再度受到关注。水力是唯一大规模用于发电的可再生能源,全世界约三分之一国家一半以上的电力依赖于水力。从全球来看,水电约占世界总电力供应的20%。大多数的大型水电站依靠大坝提供可靠的水流来驱动涡轮机,但小型"径流式"水电站不需要此类障碍物阻止河流的自然流动。雨量充沛、地形多山的国家都把水电作为主要的供电方式。挪威是一个有趣的示例。河流提供的水电除满足其自身需求外尚有盈余,因此该国已成为水电出口国。

贸易与运输

河流中的水流还有另外一种明显的功用,即为旅行、贸易和运输提供渠道。世界上许多重要城市都是依赖通航河流发展起来的,因为河流为通往内陆提供了通道,很多情况下还为通往海外提供了通道。泰晤士河和伦敦就是很好的示例。在中世纪的英格兰,沿河货物运输在伦敦的城市发展中发挥了重要作用,实际上泰晤士河流域其他许多定居点的情况也是如此。水运在当时很有吸引力,因为成本相对较低,陆路运输的谷物和羊毛等商品的价格可能是那些水路运输的十倍以上。廉价的水路运输刺激了经济发展,市场规模不断扩大,区域专业化日渐形成,城市化持续推进。1300年左右的泰晤士河干支流运输史研究表明，水路极大地拓展了为首都供应粮食和燃料的市场。当时伦敦周边地区的农业专业化也可能是水路运输发展的结果，因为有些地区比其他地区更适合生产特定作物。水路运输对城市发展的主要影响有两个方面。对伦敦来说，廉价河流运输网的发展降低了首都食品和燃料的成本，打破了城市扩张的限制。水路运输也刺激了伦敦以外的首都腹地的城市发展，泰晤士河畔的亨利镇等城镇发展成为向城市供应农产品的专业中心。

通航河流也成为贸易的主要动脉，并刺激了中世纪英格兰其他地区更大定居点的发展。格洛斯特和布里斯托尔依托塞文河进行水上运输，约克在乌斯河上设有码头，诺里奇在文瑟姆河上设有码头。水运对城市发展的重要性，甚至在12世纪早期盎格鲁-撒克逊国王忏悔者爱德华颁布的法律中也得到了体现。法律明文规定应当维持主要河流的通航，"船只可以从各地沿河流向城市或集镇运输货物"。

许多经济史学家认为，在英格兰，一直到18世纪，河流运输都是大宗货物最廉价的内陆运输方式。尽管如此，在中世纪，依赖河流进行贸易的船夫和商人却必须与那些想要建造磨坊和鱼堰的人士进行不断的斗争。18世纪中期被认为是英国"运河时代"的诞生期，当时实业家们纷纷建造自己的运河，这一时代延续了150年，随着英国许多河流条件的"改善"，水路运输变得越来越容易。

河流运输在许多国家的经济发展中发挥了关键作用。以瑞典为例，在17、18世纪，由北部森林砍伐下来的木材顺流而下一直漂流至中部的矿区，被用作冶炼作业的燃料。19世纪下半叶，

河流

由于以锯木厂和纸浆厂为基础的出口导向型林业快速发展，这种形式的河流运输在瑞典的工业化进程中发挥了重要作用。瑞典北部偏远地区的森林能够满足发展中的西欧工业经济体对锯材和方木日益增长的需求。瑞典的主要河流及其支流通常由北向南流动，密集的河网为沿海锯木厂和纸浆厂所需木材的廉价长途运输提供了可能，所以采伐北部大片地区的木材是可行的。瑞典明显的季节性气候也有利于木材的运输，春季融雪使河水 65 上涨，方便了木材的漂流。20世纪初，瑞典出口总值中大约一半来自锯材、纸浆和纸张。1980年代，随着公路网络的扩大，木材水上运输被放弃，河流在瑞典木材运输方面的重要性才逐渐减弱。

在孟加拉国，货物和人员的水路运输目前仍然具有重要的经济意义，约700条河流及其主要支流构成的河网是世界上最大的内陆水路运输网络之一。在旱季，现代机械化船只通航河流的总长度会有所缩短，但它们依然连接着几乎所有的主要城市、城镇和商业中心。事实上，孟加拉国的内陆港口处理了约40%的对外贸易货物。

内陆水运比公路或铁路运输便宜，而且往往是服务于农村贫困人口的唯一方式，在季风季节大范围洪水期间许多道路无法通行时水路运输特别有用。数百年来，乡村船只是孟加拉国传统的水路运输手段，也是公路网络不发达的南部地区主要的全天候运输工具。

在世界上的某些地区，沿河贸易中截然不同的文化之间的接触带来了一系列有益或有害的影响。在北美洲，受毛皮贸易的刺激，欧美于18世纪开展密苏里河谷探险。美洲土著印第安

图9　几十万艘乡村船只在孟加拉国的河流上穿梭，运送乘客和货物。这些船只在农村的生活和经济中发挥了至关重要的作用

人在沿河的某些地方接触欧洲商人，其中一些贸易中心早在几百年前就已经与欧洲人接触。美洲印第安人用海狸皮和野牛皮换取金属炊具、刀、枪、织物、珠子、咖啡和糖等制成品和加工品。19世纪，来自商贸基地林立的圣路易斯镇的蒸汽船在密苏里河上往返穿梭。美洲印第安人接触了欧美文化的许多方面，但也在无意中接触感染了他们没有免疫力的致命疾病。1837年，很可能是一名汽船乘客导致了平原印第安部落的天花流行，致使1万～2万名印第安人死亡，其中作为密苏里河一个主要贸易站所有者的曼丹族，90%以上的族人在此次天花流行中丧生。

多瑙河：欧洲的大动脉

尽管多瑙河不是欧洲最长的河流，但许多人还是将其看作

66

欧洲最重要的河流，就像17世纪中叶教皇英诺森十世批准在罗马纳沃纳广场修建四河喷泉时一样。四河喷泉是吉安·洛伦佐·贝尔尼尼最引人注目、最壮观的作品，它由四座大理石雕像67组成并以埃及方尖碑为顶，这些雕像象征着当时已知的世界主要河流（毫无疑问，部分程度上指的是伊甸园中的四条河流）。尼罗河代表非洲，恒河代表亚洲，拉普拉塔河代表美洲，多瑙河代表欧洲。多瑙河是世界上连接国家最多的河流，它定义并整合了整个欧洲大陆。

人类在多瑙河流域活动的历史至少可以追溯到25 000年前，当时人们聚集在今天的捷克共和国的下维斯特尼采狩猎猛犸。作为连接欧洲大陆东西部的天然迁徙通道，大约7 000年前，安纳托利亚半岛的农民就利用这一通道寻找新的可耕地。5 000年后，波斯国王大流士率领庞大的军队沿着同样的路线前进并渡过多瑙河攻打斯基泰人。多瑙河是古希腊人建立的贸易走廊，在罗马时代，多瑙河还兼具防御屏障，以及为沿岸驻扎的军团士兵提供食品和装备的补给线的双重功能。

1 000年前，在十字军东征时期，多瑙河是欧洲基督教军队开往拜占庭和圣地的通道；16世纪，多瑙河成为一条反向的十字军之路，伊斯兰教在此时被苏莱曼大帝从黑海带到了西方。1520年代，奥斯曼土耳其人占领了贝尔格莱德，打败了匈牙利人并向维也纳城墙挺进。他们在占领布达佩斯150年后才被赶回多瑙河。

多瑙河沿岸贸易造就了奥地利和匈牙利两大帝国，这两个帝国曾并于哈布斯堡王朝之下，德语人士称之为多瑙河王朝或"多瑙河君主国"。奥地利女大公、匈牙利和波希米亚女王玛丽亚·特蕾莎专门成立了一个帝国政府部门监管多瑙河航运。今天，曾被拿破仑·波拿巴昵称为"欧洲河流王子"的多瑙河流经欧洲大陆的四个首都城市（维也纳、布拉迪斯拉发、布达佩斯和贝尔格莱德），并跨越或流经十个国家的边界。作为欧洲动脉的多瑙河一直在贸易中发挥重要作用，19世纪第一个多瑙河委员会成立后，整条多瑙河的航运得到了促进。1998年成立的保护多瑙河国际委员会致力于确保整个多瑙河流域淡水资源的可持续和公平利用，包括改善水质与建立防洪和事故控制机制。

鉴于多瑙河在欧洲历史上发挥的重要作用，它反映在欧洲文化的各个方面并不令人意外。在贝尔尼尼的罗马四河喷泉出现之前，多瑙河已于16世纪孕育出了风景画流派。大约200年后，多瑙河成为小约翰·施特劳斯一首著名华尔兹舞曲的主题。河流在诸多方面给予作家和艺术家以激励与灵感，上面列出的只是其中的一些示例，下一章将详细探讨这一主题。

前进的道路

　　河流就是前进着的道路，它把人们带到他们想要去的地方。

　　　　　　　　　　　　　布莱兹·帕斯卡（1623—1662）

　　　　　　　　　　　　　　（法国数学家和哲学家）

　　千百年来，河流一直吸引着人类。它们在文化的诸多方面发挥了重要作用，为包括诗人和音乐家在内的各类文艺工作者提供了流动的灵感。河流的水流不仅被用来体现大自然田园风光的神秘，也被用来承载思想和主题，推动作家们走进过去的岁月。作为上帝作品的永恒象征，河流将精神和物质融为一体，让人们洞察人类在事物秩序中的位置。从维吉尔的诗歌到弗朗西斯·福特·科波拉的电影，河流对文学和艺术的重要性源远流长。

河流与语言

　　河流与文化的联系悠久而丰富，其中蕴含着许多有趣的语言内涵。许多河流的名称本身就是描述性的。某些大河因其水流规模令人惊叹而被直接称为"大"或"巨大",如名称源于阿尔冈琴语的加拿大渥太华河。而另外一些河流的名称则更为形象一些。在英格兰,泰晤士河(Thames)的名字被认为源于印欧语,意为"暗河";威洛河(Wellow)蜿蜒曲折,斯威夫特河(Swift)水流湍急,克雷河(Cray)清澈纯净。在英国,源于凯尔特语的河流名称比比皆是:达特河(Dart)的名字在凯尔特语中意为"生长着橡树的河流",而艾韦恩河(Iwerne)的名字被认为是表示"两旁都是紫杉"。相反,许多河流的名称仅表示"河流"。英格兰西部埃文河(Awn)的名称源于凯尔特语,意为"河流",所以埃文河的字面意思是"河河"。类似地,南亚恒河(Ganges)的名字来自梵语单词*ganga*,意为"水流"或"河流"。

河流的名字也被用作地名。以河流名称命名的城市包括俄罗斯首都(莫斯科:莫斯科河)、立陶宛首都(维尔纽斯:维尔尼亚河)、中非共和国首都(班吉:乌班吉河)和马拉维首都(利隆圭:利隆圭河)。伯利兹首都贝尔莫潘以该国最长的伯利兹河及其支流莫潘河的名字命名。在更大的尺度上,许多国家以其主要河流的名称命名。这些国家包括南美的巴拉圭、中东的约旦、西非的冈比亚和塞内加尔。在西非东部,尼日尔河流经尼日尔和尼日利亚,中部非洲的刚果河孕育了刚果共和国和刚果民主共和国。印度以印度河命名,尽管它已不再流经印度。1806年,拿破仑·波拿巴在北欧建立了一个类似国家,即莱茵河联邦,但1814年拿破仑退位后联邦瓦解。

同样,许多地名也与河流存在着间接的联系。牛津的意思

河流

是"牛穿过的地方"或"浅滩"。剑桥可以追溯到"格兰塔河上的桥",格兰塔这一河名源于凯尔特语,后称为剑河被认为是受诺曼语的影响。许多位于河口的定居点的名字其词源也同样简单:雅茅斯和法尔茅斯分别位于雅尔河和法尔河的河口。当然,这一原则也适用于其他许多语言。阿伯丁(Aberdeen)是苏格兰东北部的一个港口,其名称来源于凯尔特语(唐河,现称迪恩河的*aber*或河口)。同样,丹麦东部港口奥尔胡斯(Aarhus)在古丹麦语中就是"河口"的意思(*aa*,河;*os*,口)。在美国,许多州名都源于美洲土著的与河流相关的词汇。康涅狄格源于莫希干语,意为"长河之地";密西西比源于齐佩瓦语,意为"大河";密苏里源于阿尔冈琴语,意为"大舟之河";内布拉斯加源于奥马哈或奥托斯印第安语,意为"宽阔的水"或"平坦的河流"。然而,并非所有与河流有关的地名都是可靠的。巴西沿海城市里约热内卢就是一个很好的示例,它由1502年元旦那天第一次发现该地的葡萄牙水手命名。他们称之为"一月河",错误地认为里约现在所处的大海湾是一条大河的河口。

　　一些源于河流的术语在英语中被广泛使用。meander(曲流)一词就是一个很好的示例,在地方语言中它被用作动词和形容词,表示蜿蜒的道路。rival这个词也源于河流术语,表示为同一目标与他人竞争的人。它来源于拉丁语单词*rivalis*,意为"使用同一条河流"。著名成语crossing the Rubicon有其历史渊源,字面意为越过卢比孔河,引申义破釜沉舟。卢比孔河是东、西罗马两国之间的边界,任何罗马将军都不得率领部队渡河南下,否则就是直接挑战帝国权威。因此,当尤利乌斯·恺撒决定渡河进军罗马时,他已破釜沉舟,踏上了一条不归路。

风景画

　　河流及其河谷为世界各地的风景画家提供了丰富的灵感源泉。曲折的河道蜿蜒穿过悠久的中国山水画历史。例如，宋朝最著名的画作可能是12世纪早期张择端创作的《清明上河图》。画卷全景式地描绘了宋朝都城汴京（今天的开封）的日常生活，作品以其对河流周边及沿河的大量人物、建筑、桥梁和船只的细节描摹闻名于世。这幅画被后世的二十多位艺术家模仿过。最新的仿作是2010年上海世博会上中国馆展出的电脑制作的动画版《清明上河图》。

　　欧洲早期风景画的某些示例可以追溯到16世纪初，当时德国和奥地利的许多艺术家都与多瑙河风景画派有关。他们的作品主要以帝国城市雷根斯堡为背景，将意大利文艺复兴后期的影响与德国哥特传统融为一体。300多年后，法国印象派的许多画家从塞纳河水面瞬息万变的色彩和光线效果中获得了灵感。他们包括奥古斯特·雷诺阿、克劳德·莫奈、爱德华·马奈和古斯塔夫·卡勒波特。莫奈选择居住在离巴黎不远的吉维尼村的河边。塞纳河在后来法国艺术家的作品中也具有重要地位，其中包括乔治·修拉最著名的点彩画作之一，1884年创作的《大碗岛的星期天下午》（大碗岛是塞纳河中的一个岛屿，当时是肮脏巴黎的田园度假胜地）。塞纳河也为野兽派画家亨利·马蒂斯和莫里斯·德·弗拉芒克提供了早期的灵感，后来他们搬到了气候较为温暖的地中海地区。

　　在欧洲其他地方，19世纪早期英国浪漫主义画家约翰·康斯太勃尔与斯托尔河有着特别密切的联系。康斯太勃尔出生于

河流

图 10 《小船》（*La Yole*），皮埃尔—奥古斯特·雷诺阿绘于 1875 年。作为画面背景的塞纳河曾对许多印象派画家产生过巨大的影响

74

东英格兰斯托尔河畔的一个小村庄东伯格霍尔特，而这条河的周边地区戴德汉姆河谷，在这位艺术家生前就以康斯太勃尔之乡闻名于世。差不多同一时期，切尔涅佐夫兄弟关于伏尔加河的作品使得人们更加欣赏俄罗斯艺术中的风景（见下文）。

在北美洲，一位名叫托马斯·科尔的艺术家于1825年第一次沿哈德逊河前往卡茨基尔，其旅行过程中创作的作品在新兴的纽约艺术界引起了轰动。由此诞生的哈德逊河画派宣称它是美国第一个连贯的艺术流派。该画派成员早期主要关注纽约州哈德逊河沿岸宏伟的自然风光，赞美原生态的景观，但后来其主题范围逐渐扩展，甚至包括遥远的南美和北极。另一位作品与河流密切相关的美国艺术家是约翰·班瓦德，他关注的是密西西比河。1840年，班瓦德开始创作密西西比河大型全景画，该油画最终长度约800米（约半英里，尽管广告上说它有3英里长）。班瓦德向付费公众展示其作品，后来他带着密西西比河全景画去了欧洲，于1849年在伦敦附近的温莎城堡为维多利亚女王做了私人展示。

伏尔加河：俄罗斯的灵魂

伏尔加河是欧洲最长的河流，它在俄罗斯人民的心目中拥有特殊地位，是他们钟爱的民族文化象征。在民间传说、歌曲、诗歌和绘画中受到尊崇的"母亲河"或"伏尔加母亲河"代表俄罗斯广袤的区域，体现了俄罗斯历史的命脉。在尼古拉·卡拉姆津、伊万·德米特列耶夫和尼古拉·涅克拉索夫等19世纪作家的感伤诗中，伏尔加河被描绘成俄罗斯的象征。19世纪上半叶号称是俄罗斯诗歌的黄金时代，其领军人物彼得·维亚泽姆斯基王子把伏尔加河誉为"民族的标志"。马克西姆·高尔基在其长篇和短篇小说中也生动地描绘了伏尔加河流域人民的生活,他早年生活在伏尔加河和奥卡河交汇处的下诺夫哥罗德市, 75 曾在伏尔加河的轮船上做过洗碗工。

表达对伏尔加河的敬意是俄罗斯民歌常见的主题,《伏尔加河船夫之歌》就是这一主题的代表作,这是蒸汽时代之前纤夫在伏尔加河某些河段上沿岸用纤绳拉船时通常会唱的一首水手号子。这首歌因出生在伏尔加河地区的歌剧男低音歌唱家费多尔·夏里亚宾的演唱而流行。它与伊利亚·列宾的生动描绘沙皇俄国农民恶劣工作条件的著名同名油画联系密切,也在涅克拉索夫的诗中获得了共鸣:"这是成群的纤夫在爬行/这呼声粗野得没法忍受。"列宾的这幅作品完成于1873年,它成功地捕捉到了纤夫们的尊严和刚毅,代表了民族现实主义画派发展的一个关键阶段。19世纪后半叶,艾萨克·列维坦、伊万·希什金和鲍里斯·库斯蒂季耶夫等著名俄罗斯艺术家越来越多地在画布上描绘了伏尔加河及其城镇、村庄和周边风光。列维坦的作品以其反映俄罗斯大自然灵魂的倾向而闻名于世。他在伏尔加河上度过了几个夏天,他的一些著名的画作捕捉到了变化的光线、生活的节奏以及伏尔加河风景的美丽和宁静。

俄罗斯乡村风景艺术获得真正欣赏可以追溯到1838年,当时,受沙皇尼古拉一世的皇宫事务部的派遣,格里戈里和尼卡诺·切尔涅佐夫兄弟俩踏上了从雷宾斯克到阿斯特拉罕的伏尔加河"发现之旅",全景绘制"伏尔加河两岸美景"。最终在圣彼得堡展出的是一幅长约600米的环形风景画,展厅装饰类似船舱,并配备了模拟水上旅行的音效。遗憾的是,这幅史诗般的作品在历经无数次翻卷后并没有保存下来,但切尔涅佐夫兄弟的日记、旅行笔记以及部分工作草图和油画保存了下来。

在电影方面,音乐喜剧《伏尔加,伏尔加》是苏联时期的一部经典影片,据说是苏联领导人约瑟夫·斯大林最喜爱的一部电影。影片讲述了一位才华横溢的民谣歌手为了去莫斯科参加音乐比赛而与小官僚斗争的故事,主要场景为伏尔加河上一艘名为"约瑟夫·斯大林号"的轮船。该片于1938年首映,其轻松的逃避现实主义风格与苏联当时发生的经济困难和政治清洗形成了鲜明的对照。

音　乐

理查德·瓦格纳的大型四联剧《尼伯龙人的指环》(英文中通常简称为《指环》)的主人公是莱茵河中的三个水精灵。莱茵河少女(借用德国民间传说中的水精灵——见第二章)是藏在莱茵河中的莱茵黄金的守护者,在19世纪中期史诗中,宝藏后被偷走并变成了戒指。她们出现在第一和最后一个场景中,最终从莱茵河水面升起,从布伦希尔德火葬柴堆的灰烬中取回戒指。

罗马尼亚作曲家扬·伊万诺维奇1880年创作的华尔兹圆舞曲《多瑙河之波》展现了多瑙河的魅力和浪漫,但14年前由奥地利指挥家和作曲家小约翰·施特劳斯创作的圆舞曲更为广受欢迎。这首德文名称为*Ander schönen blauen Donau*,在英语世界更为熟知的名字是《蓝色多瑙河》的圆舞曲,从那时起便一直是极受欢迎的古典音乐作品。

约翰·施特劳斯所生活和工作的维也纳是当时的奥匈帝国首都,也是高雅文化和古典音乐的中心。在当时的帝国行省波希米亚，捷克作曲家贝德里赫·斯美塔那创作了民族交响诗套曲《我的祖国》，其中的《伏尔塔瓦河》是至今仍最受欢迎的片段。音乐描绘了贯穿波希米亚的伏尔塔瓦河流过森林和草地，经过废弃的城堡，见证农民的婚礼，然后壮阔地穿过布拉格汇入易北河。这首令人回味无穷的作品奠定了斯美塔那作为捷克民族乐派运动奠基人之一的地位，许多人认为《伏尔塔瓦河》是捷克共和国的非官方国歌。

密西西比河是另外一条有着深厚音乐传统的河流，尤其是在它流经的被称为美国南方腹地的下游地区，在19世纪和20世纪的大部分时间里，该地区是一个文化凝聚力很强的农业区，主要种植棉花。起源于密西西比河这一区域的各种音乐风格在整个北美洲及其他地区受到欢迎。蓝调诞生于密西西比河三角洲，后者为密西西比河和亚祖河之间的冲积洪泛平原，而布吉伍吉和爵士乐则诞生于下游的新奥尔良。蓝调与福音音乐融合产生了节奏蓝调、摇滚乐和灵魂乐。路易斯·阿姆斯特朗、B.B.金、查克·贝里、杰瑞·李·刘易斯、埃尔维斯·普雷斯利和艾瑞莎·弗兰克林都是密西西比河畔出生并成长的20世纪国际著名音乐家。

文学中的河流

作家和诗人以多种方式利用河流。河流不仅可以作为一种地理特征，也可以作为一种文学手法，其恒久的运动和不变的方向为叙事提供了动力。河流之旅是最常见的河流隐喻之一，它将过去与现在联系在一起，还兼作生命之旅，呈现对成长经历的洞察。作为小说的背景，河岸提供了一种命运感，暗示了自我发现的可能性。

通过对罗马文学中河流被用作诗歌手法的各种方式的评估，人们可以清晰地看出这种充满活力和变化的景观元素是如何与诗歌的动力相互作用的。河流可以是诗歌和诗人之间的媒介，河流的流动可以是叙事的一部分，也可以是叙事结构的一部分。尤其是在维吉尔的史诗《埃涅阿斯纪》（前19）中，河流象征着方向性的进步，旅程同时具有空间性、时间性和文学性。台伯河既是埃涅阿斯意大利之旅的起点，也为叙事提供了方向。

河流推动诗歌叙事的另一个示例是阿尔弗雷德·丁尼生的诗歌《夏洛特夫人》（1833）。诗中的一切都随着河水的流动而变化。当夏洛特坐在塔楼上时，河水从她身边流过，映照着这个世界，向下游的卡米洛特流去。当兰斯洛特爵士骑马匆匆而过时，夏洛特离开塔楼进入河中的真实世界，她解开河岸上的船链并在船头上写下了自己的名字，通过确立身份有效地发现了自我。她随船顺流而下到达卡米洛特，死在了那里。

马克·吐温的《哈克贝利·费恩历险记》（1885）是一个典型的以密西西比河为背景的河流故事，其主题清晰地呈现了自由、变化和蜕变这些大河河道所固有的特质。因不堪酗酒成性的父亲的虐待，哈克·费恩和他的朋友吉姆（一个逃跑的奴隶）乘着木筏沿密西西比河逃走，他们的旅程代表了对压迫、破碎的家庭生活、种族歧视和社会不公的逃离。该书取材于作者在密西西比河上的童年经历。塞缪尔·克莱门斯（马克·吐温的真名）二十多岁时曾做过内河引航员，其笔名就源于这段经历，取用的是观测员测量河道浅水水深时常见的叫声。为避免船只搁浅，观测员要向引航员报告水深，"马克·吐温"意为

"测标2英寻"。

在查尔斯·金斯利的经典儿童小说《水孩子》（1863）中，这种变化和新生更具奇幻色彩，小说以扫烟囱的男孩汤姆探求河流的清洁功能开头。汤姆逃离可怕的生活去河中寻找自由，但在经历了水孩子的历险之后，他最终在基督教救赎的道德故事中再次以人类的形态重生。围绕都市河流展开的小说《我们共同的朋友》是查尔斯·狄更斯最具影响力的作品之一，该书于金斯利的《水孩子》出版一年后开始创作。小说以连载形式出版，作品利用维多利亚时代的伦敦泰晤士河赋予若干人物以重生和新生，充满了水的意象。威廉·博伊德在《普通雷暴》（2009）一书中也以类似的方式利用泰晤士河来改变身份。当时，伦敦警察平均每周都会从泰晤士河里捞出一具尸体，这一骇人听闻的事件促成了该书的创作。

在文学作品中，借由其对边界或阈值的表征，河流也可以用作转变的催化剂，所以渡河行为会促成某种改变。河流可合可分，可以是伙伴，也可以是神。因其蕴含着大自然的核心奥秘，河流可体现对智慧的追求。我们也可以借助它们来探索物质世界，寻求道德、知识与自然法则的方向。当然，即使在一部作品中，河流也可以有很多含义。

刚果河:《黑暗之心》

许多人认为，约瑟夫·康拉德的《黑暗之心》是最重要的"现代主义"小说，这部作品极具复杂性，旨在反映现实世界中的复杂感受。非洲的刚果河作为贯穿全书的主线，为小说内容

80

的不确定性赋予了方向和形式。故事描述了一个并不复杂的任务，即马洛去河流上游冒险寻找库尔茨。这是一次身体上的旅行，溯河而上进入非洲大陆，同时也是一次直面殖民主义残酷现实的道德和政治之旅（库尔茨是一名失踪的贸易代表，为一家从事象牙贸易的比利时公司工作）。它还是在另一个层面上进行的旅程，即一场由马洛和读者共同进行的心理之旅，在这段旅程中，我们深入自我，直面最基本的动力和冲动、弱点和需求，深入地下世界，也就是"黑暗之心"。

小说以嵌套故事的方式构建，叙述从泰晤士河的河口开始，四个男人坐在一艘船的甲板上，听马洛讲述他年轻时去非洲的一次旅行。在这样的场景设置下，发生在遥远大陆"黑暗之地"的事情，得以在听众看似安全而舒适的世界里产生反响。

在马洛向上游航行的过程中，库尔茨的形象逐渐浮现。这个作为善的力量启程的人已经被权力腐化。库尔茨在当地的非洲社区获得了近乎神的地位，这种地位因其使用武力而得到巩固：他掠夺农村地区的象牙，随意射杀百姓，把他们的头骨展示在栅栏上以显示他的权威。马洛的非洲之心旅程探索了欧洲启蒙运动、理性语言和帝国主义言论的阴暗面。

19世纪末，康拉德的《黑暗之心》首次在《布莱克伍德杂志》上连载，1902年以小说的形式面世。20世纪末，马洛的河流之旅在另外一部经典作品中再现，这次是一部电影，即弗朗西斯·福特·科波拉执导的越战大片《现代启示录》（1979）。当然，从表面上看，电影以另一个大陆为背景，但它表明，在作品问世近一个世纪后，康拉德的故事依然在当代产生了无数回响。马龙·白兰度饰演的特种部队指挥官库尔茨上校在权力中陷入

图 11 约瑟夫·康拉德的经典的经典现代主义小说《黑暗之心》使刚果河名垂不朽，但灵也有人认为，该书使西方人产生了很多很多轻视轻视哈拉哈拉以南非洲的看法[82]

疯狂，象征着启蒙运动、人道主义和所谓进步中的堕落声音。和小说一样，电影中塑造的形象和人物可以被解读为对战争、种族主义和殖民主义的强烈批判。然而，无论是小说还是电影，也都被看成这些作品试图揭露的虚伪价值观的表达。

科波拉的电影同样以嵌套故事的方式展开，同期拍摄的纪录片《黑暗之心》记录了《现代启示录》的制作过程，见证了现实生活中的腐败、堕落和疯狂，堪比虚构的库尔茨。无论如何，小说、电影以及关于电影的电影都只是站在局外人的角度讲述故事的。对于这条河流流经的陌生大陆，没有人去试图了解。人们可以批评说，这是欧洲把非洲整体尤其是刚果河，以及美国为自由和民主而进行的狭隘圣战神话化的象征。但是，对于每一个故事的多层面目标而言，可替代参照系的缺位同样至关重要。每个故事都是一次本质上孤独的旅程，包含着旅行者深刻的精神变化；这一旅程担负着深入事物核心的使命，却无法揭示关于人类生存问题的简单答案。如同他所居住的黑暗世界，库尔茨的性格神秘莫测。在所有的情形下，河流的作用都至关重要，康拉德将其称为"人类梦想"和"帝国萌芽"的通道。

河流

83

驯服的河流

> 对河流的制服是人类在征服狂暴的大自然的斗争中取得的最伟大、最重要的胜利。
>
> 爱德华·吉本（1737—1794）
>
> （英国历史学家）

在人类历史上，人与河流一直在相互作用，人类对河流的影响既有直接的，也有间接的，其影响形式多种多样。从河流中大规模抽水进行作物灌溉的最早例子可以追溯到6 000年前。利用堤坝、引水、渠化和涵洞等工程对河道进行人工控制的历史也很悠久。中东地区的一些世界最古老的水坝建于4 500多年前，而中国黄河的人工调水和治理则始于2 000多年前。自这些早期示例之后，人类对全球河流进行人工改造的范围、抱负和规模不断增大。然而，河流改造的程度和强度仍然存在明显的区域差异。在今日欧洲，由于饮用水供应、水力发电、防洪或其他方面的原因，其主要河流总流量的近80%受到流量调节措

图12 即使在人迹罕至的地区，河流也在一定程度上受到了人类活动的影响，如位于以偏僻而闻名于世的巴拿马达连地连地垫区域的河流

施的影响。个别国家的比例甚至更高。英国河流的调节比例 84 约90%，而在荷兰，这一比例接近100%。相反，其他大陆上的包括亚马孙河和刚果河在内的一些超大河流，几乎未经任何人工调节。

除了直接的、有目的的改造外，作为一种计划外的副作用，土地利用和土地利用变化对河流的影响常常也会导致河流的改变。森林砍伐、植树造林、土地排水、农业以及明火作业都会产生重大影响，其中工程建设和城市化的影响可能最严重。这些影响多种多样且并非都是直接的。动态河道及其关联的生态系统在许多方面都是相互适应的，因此，景观中的人类活动会影响水沙供应，并可能引发一系列复杂的其他变化。各种人类活动导致了当代气候变化，也在某种程度上改变了河流，许多权威人士认为，几乎没有不受人类影响的河流（即使是在世界上人口最稀少的地区）。因此，河流的演变和开发在诸多方面受到社会经济因素和自然因素的双重驱动。

灌溉农业

从依赖狩猎和野生食物采集的生存方式，向主要依靠以植物栽培和动物饲养为来源的粮食生产的生活方式的转变，是人类社会最重要的发展之一。在第三章中，我们已经讨论了早期农业管理与底格里斯-幼发拉底河、尼罗河和印度河冲积河谷沿线等世界上几个独立中心出现的城市文明之间的联系。这些联系伴随着管理永久农田和灌溉系统所需的高水平组织的形成而出现。秘鲁安第斯山脉西侧的赞拿河谷是早期利用河水进行农作物灌溉的另一个中心，考古学家在那里发现了至少5 400年 86

前,很可能是 6 500 年前使用过的小型自流运河系统。

可以说,灌溉农业今日之重要性并不亚于早期文明时期,尽管目前农田灌溉所利用的淡水包括了地下水、湖泊、地表径流以及各种形式的废水等多种水源,但河流仍然是最重要的。除一些当代工程使用水泵进行远距离配水外,从最早的河流灌溉方案产生开始,(水库)蓄水和(运河)配水的方法就没有发生过根本性的改变。无论如何,许多灌渠利用的依然是重力。世界上有一半的大坝[指15米(含)以上的水坝]是专门或者主要为了灌溉而兴建的,全球大约三分之一的灌溉农田依赖水库蓄水。在包括人口众多的印度和中国在内的一些国家中,大坝为50%以上的可耕地提供了灌溉用水。

抽取河水进行农作物灌溉的连锁效应是惊人的。在某些情况下,它可能导致河流规模、型式和形状的彻底改变。这种“河流变形”的一个示例出现在美国西部大平原,19世纪末,欧裔美国人将这里的河流描述为宽、浅的辫状河道,河岸上只有稀疏的植被,但此后这里的河流发生了巨大变化。为满足农业灌溉需要而进行的河流流量调节导致季节性洪峰流量减小、基流量增加以及区域地下水位变化,地下水的抬升促进了沿岸树木的生长。在水流情势变化和河岸阻力的共同作用下,在仅仅几十年的时间里河流就变得狭窄、蜿蜒,而河流两岸也已森林茂密。

遗憾的是,许多灌溉方案管理不善,灌区内外常常发生很多环境问题。在许多大型灌渠网络中,真正有益于农作物的水不及河流或水库引水的一半。大量的水通过无衬砌渠道渗出或者在到达农田之前就已蒸发。由于灌溉过量或施灌时机不当,一部分水也会从农田中白白流失或经土壤下渗,未被植物利用。

河
流

87

这些水大部分渗回附近的河流或汇入地下含水层,因此可以再次使用,但如果水被盐、化肥或杀虫剂污染,则可能导致水质恶化。过量灌溉往往会导致地下水位抬升,造成盐碱化和渍水。这些过程导致世界各地灌溉计划中的作物产量下降。

许多此类难题一直困扰着中亚土库曼斯坦和乌兹别克斯坦的农民,沙漠条件意味着这里90%以上的农业生产依赖阿姆河和锡尔河的灌溉。在1950年代的苏联时期,中亚地区农业灌溉的迅速发展导致了一些令人瞠目的后果。到1980年代,灌溉面积达约700万公顷,增加了一倍多。结果,为咸海提供90%水源的阿姆河和锡尔河每年排入咸海的水量下降了一个数量级,由每年约55立方千米降到每年约5立方千米。

因此,咸海毫无意外地大幅缩小。1960年,它是世界第四大湖泊,但之后其水面面积减少一半以上,体积减少三分之二,水位下降25米多。在某些区域,咸海剩余水体的含盐量是公海海水的两倍多。由于无法在咸水中生存,咸海中的大部分本地鱼类和其他水生物种已经消失,这意味着曾经占有重要地位的商业捕捞业的终结。海平面的下降对气候也产生了局部影响,裸露的海床已经成为在咸海海岸线边几百千米范围内农田上空肆虐的沙尘暴的源头。这些细小的沙尘含有盐分,加剧了灌溉农业问题。沙尘也被认为对人类健康具有破坏性影响。

对鱼的影响

人类已经在很长时间里对河流生物产生了直接影响。鲤鱼在欧洲各国河流中均可见到,但是这种鱼原先只是多瑙河及其一些支流中的本地鱼种。大约2 000年前,潘诺尼亚行省境内的多瑙河是当时罗马帝国的北部边界,当驻扎在多瑙河沿岸的大批军团士兵习惯了食用这种野生鲤鱼后,罗马人将其引入欧洲的许多河流。

鲤鱼就这样成为法国的塞纳河第一个被引入的物种。在中世纪,包括从贵族和宗教团体养鱼池中逃出的丁鱥和赤睛鱼在内的其他物种随后也被引入。19世纪后期,更多外来入侵物种(软口鱼和白梭吻鲈)经运河从更远的东部河流来到塞纳河。19世纪末,虹鳟、黑鲈、驼背太阳鱼和黑鲴等北美物种被人工引入。

20世纪,沿河修建的许多堰和船闸使洄游物种无法到达上游产卵地,本地鱼类开始从塞纳河中消失。除了鳗鱼外,塞纳河中所有的其他洄游物种——鲟鱼、鲑鱼、海七鳃鳗、海鳟、胡瓜鱼和鲥鱼,全部灭绝。塞纳河中的原有鱼类种群大概有30个左右。今天,塞纳河中有46个物种,但只有24个是本地物种。

人类对塞纳河鱼类生物的种种影响,在世界上经济较发达地区的许多河流中也相当典型。生物入侵被普遍认为是世界各地河流和其他生态系统的生物多样性的主要威胁之一。一项对覆盖全球80%以上大陆的1 000多个流域的全球淡水鱼类入侵模式的研究表明,欧洲西部和南部是全球六大生物入侵热点之一,非本地物种占每个流域物种总数的四分之一以上。这些热点地区的受威胁鱼类物种的比例也是最高的。

人们发现,人类活动的影响,特别是特定流域的经济活动水平(用GDP表示),是这一结果最重要的决定因素。这一发现或许可以从几个方面进行解释。经济繁荣的地区更容易受到生境干扰(例如大坝和水库改变了河流流量),我们知道,生境干扰是有助于非本地物种生长的。高速经济活动也有可能通过水产养殖、垂钓和观赏鱼贸易增加物种入侵的机会。与经济发展相关的产品的进口需求增加，也加大了进口过程中无意引入物种的可能性。

当然，人类活动的多重影响也是世界上较贫穷地区河流生态发生变化的原因之一。和许多岛屿一样，马达加斯加岛拥有包括鱼类在内的多种"特有"（别处没有的）物种，其淡水物种被认为极其易危。在马达加斯加特有的64种淡水鱼中，4种恐怕已经灭绝；因面临森林砍伐导致的栖息地退化、过度捕捞以及与外来物种的相互作用这三种主要压力，另外38种也濒临灭绝。

马达加斯加普遍的森林砍伐是水生生境发生多方面退化的原因之一。河流沿岸树木的减少会导致河流物种的变化，因为树木的减少意味着植物物质和昆虫掉落数量的减少，而这些掉落物正是某些鱼类的食物。此外，河岸上的树木越少，树荫也越少。更多的阳光照射使得河流水温升高，促进了藻类的生长。当以掉落物为食的鱼被那些能够以藻类为食的鱼淘汰时，物种就会发生变化。森林砍伐通常也会导致更多的径流并加剧土壤侵蚀。侵蚀产生的泥沙可能会覆盖产卵地，导致繁殖率降低。更多的泥沙还会堵塞鱼鳃，使鱼承受更大的压力，加之其他压力，鱼类可能因此而死亡。

人口快速增长导致对鱼类的需求不断增加，而环境执法又面临巨大的组织困难，因此马达加斯加的过度捕捞淡水物种问题非常棘手。该岛引入的外来鱼类包括养殖鱼类和观赏鱼类，它们对水生生态系统的影响是深远的。一些外来鱼类已经归化，完全取代了马达加斯加中部高地的本地鱼类，并在该岛上的其他地区广泛分布。

河流治理

控制河流水位和流量变化以满足社会需求的努力可以追溯到最早的文明。如今,河流治理主要是为了提供稳定的流量,以满足生活、农业和工业用水需求,以及水力发电、航行、防洪等方面的需求。河流治理的主要方法包括建造大坝(见下文),修建堰、闸等径流式水库,以及渠化。渠化是指包括拓宽、浚深、裁弯取直和稳定岸坡在内的各种河道工程措施。

最早的一些河道治理科学原理创立于意大利,人们认为莱昂纳多·达·芬奇在这里发明了利用两个内置垂直闸门控制河流水位变化的双门船闸。纳维利奥格兰德运河连接着提契诺和米兰,15世纪末,他为该运河设计的船闸极大地推动了内陆航运的发展。200年后,随着1694年博洛尼亚大学设立"水文测验学"教授职位以及一系列有关河流水力学书籍的出版,"水科学"在意大利北部成功创立。当时有人主张,辫状河的最好治理手段是将其变为单一河道,到19世纪末,西欧大多数辫状河都以这种方式进行了治理。

19世纪在欧洲出现的河流工程的另一个重要阶段,是对主要河流进行广泛的河道取直和河床加深。法国塞纳河和多瑙河三角洲的支流苏利纳河上的重大工程就属于这类性质,但最引人注目的方案之一是在流经匈牙利的多瑙河支流蒂萨河上实施的工程。为加强农田排水并减少匈牙利平原洪水泛滥,蒂萨河治理工程对100多个曲流进行裁弯取直,河流长度因此缩短了近400千米。

91

河
流

图13 1952年,流经英格兰西南部林茅斯的西林恩河下游发生毁灭性洪
水,34人在洪水中丧生。洪水过后,下游河道实施了渠化措施,拓宽了河
道,加固了堤防   92

黄　河

　　很多世纪以来,黄河一直是最引人注目的河道管理的历史案例之一。尽管在中国仅是排名长江之后的第二长河,它却是世界第四长河,同时也被认为是世界上含沙量最大的河流,在流经黄土高原后进入华北平原,每年挟带的细黄沙达16亿吨,因而得名黄河。黄河发源于青藏高原,流经5 000多千米后注入北太平洋的一个海湾——渤海。但它并非一直如此。像许多河流一样,黄河多年来也历经了改道,不过其改道比大多数河流更加频繁。事实上,在过去的大约2 500年里,平均几乎每个世纪黄河就会发生一次重大改道。有时,它不是流入渤海,而是流入向南300多千米的黄海。有几百年的时间,它根本没有流入大海,而是流入湖泊。

　　在人口稠密的中国东部平原上,河流的每一次改道都意味着一场重大的洪涝灾害。事实上,黄河的洪水习性也让它获得了"中国的忧伤"的别称。1642年9月,具有相当规模的城市开封附近的一场洪水淹死了大约34万人,留下的幸存者只有3万。早在2 200多年前,中国人就开始在黄河两岸修建堤坝,试图防止此类洪水。21世纪初,黄河下游至入海口的870千米长河堤修建完毕。河堤修建很可能挽救了许多人的生命,但多年以来,黄河大堤多处决口仍然导致了灾难性的洪水泛滥。

　　其中1938年的大堤决口是人为的。抗日战争期间,中国国民政府下令军队炸毁花园口的黄河大堤,企图利用洪水阻止日军前进的步伐。虽然有几千名日军被淹死,但洪水只是拖延了日军前进的速度。当地的中国百姓成了这场灾难首当其冲的受

河
流

害者，11座城市和4 000多个村庄被洪水淹没，共约1 200万人受洪水之害，其中近90万人被淹死。九年后工程技术人员才修复了花园口大堤，黄河重新流入渤海。

数百年来的堤防建设也产生了其他影响。大多数河流的下游都会沉积淤泥和泥沙，黄河也不例外。然而，由于修建了大堤，黄河下游很少发生洪水，大部分淤泥和泥沙都沉积在河道自身的河床上。因此，几个世纪以来，河道的高度慢慢上升，大堤也不得不相应加高。今天，下游的河床比堤外的地面平均高出5米左右。在开封，河床比街道地面高出13米。新乡市的居民生活在黄河下方至少20米处。出现这种现象的河流常被称为"悬河"。

1960年代以来，黄河中上游地区相继修建了一批大型水坝和水库。这些工程有助于防洪，同时也为依赖这条河流生存的一亿居民提供了淡水。对黄河水量日益增长的需求造成了水资源短缺，以至于1990年代初的一段时间内黄河无法入海。1997年，黄河"断流"天数达226天，有时断流位置向内陆延伸达700千米。从那时起，中国政府一直确保黄河入海，尽管水量不大。但黄河目前每年向北太平洋输送的泥沙肯定远远少于10亿吨。由于黄河悬河河段的水量实际上很小，因此发生洪水的可能性降低，但依然有可能的是，上游发生大坝无法抵御的大洪水，下游大堤再次溃决并带来不堪设想的后果。

大　坝

筑坝是人类改变河流最深刻的方式之一。以这种方式拦截河流并控制流量会带来一系列的变化。大坝拦截泥沙和养分，改变河流水温和化学成分并影响侵蚀和沉积过程,河流因此塑造景观。大坝通常通过削减洪峰流量和增大最小流量使流量变得更加均匀。流量的自然变化对河流生态系统及其生物多样性都很重要,大坝使流量变得均匀,其后果通常是鱼类物种和数量减少。

尽管人类在河流上建造大坝的历史已有数千年,但随着土方工程和混凝土技术的进步,在过去的50多年里,世界各地的大坝建设速度和规模都明显增加。21世纪初,全世界约有大坝80万座,一些大坝的高度超过200米。某些河流被人类以这种方式进行了集中的调节。例如北美洲的哥伦比亚河,19世纪中期以来,这条河流上的大坝已经不下80座。在某些大型水系中,大坝的蓄水容量超过了河流的年平均径流量。西非沃尔特河水库的蓄水容量超过河流年平均径流量的四倍。一般认为,世界上的主要水库控制了大约15%的全球地表径流。目前全世界各种规模的水库中的蓄水总量不少于全球河流年径流总量的五倍,人们认为,这种巨大的水量再分配导致地球轨道特征发生了很小但可观测的变化。

人类修建大坝的初衷是防洪以及提供农作物灌溉用水和生活用水。现代大坝仍然具备这些功能,但增加了水力发电和工业供水等其他许多功能。毫无疑问,许多大坝方案非常成功地实现了它们的目标,并在许多方面为河流资源的可持续利用做出了重大贡献。在埃及,阿斯旺大坝自1970年竣工以来一直被视为经济发展和国家声望的重要象征。其发电量占全国电力的20%,坝后水库纳赛尔湖的蓄水使灌溉农业得以发展,灌溉农田面积增加5 000平方千米。这对于一个可耕种面积很小的沙漠

96

国家来说尤为重要。纳赛尔湖的建成还催生了新的捕捞业。该大坝具有防洪抗旱功能，能够调节流量巨大的季节性变化，使尼罗河流量趋于均匀。河道水位的稳定也有利于航运和旅游业的发展。

尽管许多大坝在实现其主要目标方面是成功的，然而，大坝及其水库的建设也导致环境发生了明显变化，其中许多变化被证明是有害的。变化的确切性质和大小取决于水库的类型及其运行方式，也取决于受影响流域的性质。新建大坝产生的最明显的影响是库区的淹没，与之相关联，水文、植被、野生动物、局部气候甚至构造过程也受到影响。

拦蓄河流所形成的水库，其水质在最初十年左右的时间内通常会发生显著变化，之后达成新的生态平衡。由于被淹没的植被和土壤释放有机结合态元素，初期蓄水时生物生产量会很高，但之后会下降。特别是在热带和亚热带地区，新建水库养分富集的一个典型影响是蓝细菌这一有毒微藻的大量繁殖。摄入足量的蓝细菌毒素对人类和动物都是有害的，会导致一系列的胃肠道和过敏性疾病。兴建大型水库的另一个生物后果是伊乐藻迅速扩散，危害航行安全并产生大量次生影响，特别是其蒸散作用造成水量大量损失。苏里南共和国布罗科蓬多水库曾经发生的夸张情形就是一个示例，在水库建成的头两年里，快速繁殖的凤眼蓝竟然覆盖了大约一半的湖面。

有些水库非常大，例如布罗科蓬多水库，其面积约 1 500 平方千米，而加纳阿科松博大坝的坝后水库沃尔特湖的面积是前者的五倍多，是世界上最大的人工湖。人们认为，如此巨大的新水体的形成会对局部气候产生影响。沃尔特湖建成之后，加纳

图14　澳大利亚悉尼附近的沃勒甘巴大坝是世界上最大的提供生活用水的大坝之一。其水库长52千米，为悉尼地区约400万人口提供了80%的生活用水

中部的降雨高峰季节从每年的10月份转变为7月/8月份。巨大蓄水量对地壳岩石产生了压力，一些特别深的水库因此可能会引发地震。塔吉克斯坦中部瓦赫什河上的努列克大坝，是有文献记载的大坝引发地震活动的最佳例证之一。尽管中亚的这部分地区本就位于构造活跃区，但在大坝使用期的前十年中，水库蓄水初期和水位大幅上升期内的地震频度都明显增加。

修建新的大坝意味着原先居住在水库规划区内的居民必须全部迁走。涉及人员数量可能巨大，其中牵涉人数最多的是中国的几个移民方案。黄河三门峡工程移民30万人；长江三峡大坝工程移民约120万人，涉及13个城市、140个城镇和1 000多个村庄。政府通常会给水库移民提供补偿，但在许多偏远地区，居民没有居住土地的正式所有权文件，这一问题可能会延缓或实际上妨碍法律补偿。

水库下游河流的水文情势会因水库的兴建而改变。流量、流速、水质和热学特性都受到影响，导致河道及其景观，河边、三角洲、河口和近海的动植物发生变化。大坝使水流速度减缓，起到了拦截泥沙的作用，河流下游输沙量因此减少。因此，大坝下游的水流具有很强的侵蚀能力。进入河流三角洲的泥沙减少使得海岸侵蚀和海水入侵加剧，导致三角洲生态系统中的盐分增加。莫桑比克卡奥拉巴萨大坝的建设使得河流下游盐度发生变化，赞比西河河口的红树林因此受到威胁。其连锁效应之一是在红树林中繁殖的明虾和褐虾数量减少。

一些已经筑坝的河流，其下游受到的影响巨大。科罗拉多河是美国最繁忙的水道之一，该河天然状态下的输沙量极大，这也是西班牙探险家弗朗西斯科·加尔斯当初将其命名为科罗拉多河的缘由（Rio Colorado 在西班牙语中意为"红色河流"），但20世纪该河上修建的一系列大坝严重削减了原本巨大的输沙量。1930年以前，该河每年向加利福尼亚湾三角洲输送一亿吨以上的悬移质泥沙，但从1964年格伦峡谷大坝建成到1981年坝后水库鲍威尔湖首次蓄满期间，该河既没有向三角洲输送泥沙，也没有向大海输水。从那以后，只有大坝泄洪时河水才会不定期地流入加利福尼亚湾。平均而言，该河现在每年向加利福尼亚湾输送的泥沙量比1930年前的平均值小了三个数量级。河流向河口和加利福尼亚湾输送淡水和养分的减少对生态产生了巨大的影响。一项研究表明，目前河流挟带的养分不足可能导致墨西哥科罗拉多河三角洲贝类种群数量下降了96%。

大坝对河流生态的影响是多方面的。生态影响的其他重要驱动因素包括河流水温变化、溶解氧含量以及大坝对植物扩散和动物迁徙的屏障作用。自中世纪以来，欧洲人就已经认识到了大坝对洄游鱼类及其产卵通道的影响。1214年苏格兰颁布的一项法令要求所有大坝都要设置鱼道，拦鱼网周六要吊起，以便鲑鱼通过。然而，可以肯定的是，目前这个问题依然存在，有时还会产生相当大的经济影响。例如，20世纪末作为鱼子酱来源的欧洲鳇的捕获量急剧下降，其主要原因是伏尔加河上修建的几座大型水电站大坝导致其产卵地丧失。

最近一些国家掀起了一场拆坝运动，原因之一就是鱼类迁移受到干扰。被拆除的大坝数量很少，包括失去使用价值、维护成本过高或环境影响程度目前被视为不可接受的大坝。已经拆除或考虑拆除的大坝大多数位于美国，但一些欧洲国家也开展了大坝退役工作。例如，根据政府制订的卢瓦尔河及其流域长

河流
100

期管理计划，即卢瓦尔河大自然计划，1998年法国炸毁了卢瓦尔河支流上的两座大坝并清理了残骸。该计划的主要目标是确保卢瓦尔河水环境的保护并恢复该河的鲑鱼种群。拆除维埃纳河上的红房大坝和阿列河上的圣艾蒂安德维冈大坝的目的是恢复鲑鱼产卵地通道。

土地利用

河流与其流经的景观关系密切，因此，景观的任何变化都会不可避免地对河流产生影响，了解这一点这并不令人意外。人类利用景观的方式在不同尺度不同方面对河流产生了强烈的影响。例如，众所周知，清除天然森林植被以增加耕地会导致降雨截留量、雨水土壤下渗量和蒸散发量的减少，以及地表径流量的增加，后者常常引起土壤侵蚀率增加，在某些情况下土壤侵蚀率甚至增加几个数量级。这些土壤大部分进入河流，导致河道形态和生态发生相关变化。这类河流变化的记录世界各地都有，最早的记录出现在数千年前的地中海和中国的农业区，最近的记录则出现在其他一些地方。其他形式的粮食生产也会导致径流增加和侵蚀加剧。放牧和牲畜践踏减少了植被覆盖，造成土壤压实，降低了土壤入渗能力。

当雨水流经集约型农业区或经其土壤下渗时，农药和肥料的残留物会被带走并随之进入河流。这样，农业已经成为世界某些地区河流的主要污染源。1950年代以来，肥料的使用使欧洲和北美洲许多河流中的硝酸盐和磷酸盐的浓度明显升高，并导致了一系列环境、社会和经济问题，这些问题被统称为"富营养化"，即营养富集导致生物生产力提高。藻类生长是主要问题，它危害人类健康,增加饮用水处理成本,并对其他河流物种产生影响。例如,在水流缓慢的河流中,藻类生长会降低透光度并消耗水中的含氧量,有时还会导致鱼类死亡。

当然,很多此类影响可以通过加强农田水土保持得到控制。这些措施针对各种不利影响,尤其是农田水土流失对作物产量的不利影响。中国在黄河流域进行的大量研究已经证明了植树和修建梯田等水土保持措施的作用,该地区实施这些措施主要是为了减少黄河沿线水库的泥沙淤积。如果恢复以前的植被覆盖,那么,停止那些加剧径流或泥沙产生的土地利用也可能减少这些影响,但这种情况并不一定发生。在秘鲁安第斯山脉中部进行的调查发现,由于缺乏农民的打理,环境过于干燥导致无法种植作物,废弃农业梯田的土壤侵蚀率增加。

采矿是另一种产生类似影响的土地利用形式。在西伯利亚西部,科累马河流域内广泛的金矿开采扰乱了植被,加剧了侵蚀,导致其输沙量在1970年代和1980年代增加了一倍以上。有趣的是,资料表明,科累马河同期流量并无显著变化趋势,这意味着径流量没有发生变化。许多采矿作业还造成了河流污染。废弃的岩石和"尾矿"(即矿石提取矿物后留下的杂质)通常仍然含有可以渗入土壤和河道的金属。1998年,西班牙西南部阿斯纳尔科利亚尔黄铁矿发生的池塘污水意外泄漏事故,对瓜迪亚马尔河和科托多尼亚纳湿地的鸟类、鱼类和其他水生物种造成了巨大伤害。泄漏的酸性污水中含有浓度对野生动物来说足以致命的砷、铅和锌。采矿业长期以来一直对河流都有影响。罗马人发明了水力采矿技术,利用大量河水将土壤和岩石击碎并冲走,使矿物露出。这些技术被广泛运用于西班牙西北部冲

河流

积矿床的黄金开采。

河流与其周围景观中的人类活动之间的大量联系，以及由此而来的流域整体管理的重要性，多个世纪以来已经得到了认识。例如，日本政府为保持河道稳定而对山区河流沿岸的木材砍伐做出管控，可以追溯到 1 200 年前。同样，为了保护农业粮食生产和鱼塘，夏威夷传统的原始公社制度包括了流域一体化管理。为了给河流下游的农田和鱼塘提供养分，高地森林受禁令保护。在现代，这种整体管理方法则体现为"流域管理规划"。在欧盟国家，流域管理规划已成为所有主要流域的强制性规定。

密西西比河

密西西比河及密苏里河的流域面积占美国本土的三分之二，受大量人类活动的影响，密西西比河在过去大约200年里发生了显著的变化。19世纪初蒸汽船问世以来，内河运输迅速发展，大量森林被砍伐充当锅炉燃料，树木减少破坏了河岸稳定，促使航道产生不可预测的迁移。密西西比河流域的森林砍伐和商品农业的扩张也导致了土壤侵蚀加剧，更多泥沙流入河流。威胁航行安全的沙洲就是在此过程中形成的。随着定居点向地势低洼的岸边扩展，密西西比河洪水的威胁更大。

从19世纪至今，系统解决密西西比河问题的努力一直没有停止过。在整个19世纪，美国陆军工程兵团在密西西比河上清除岩石，浚深特定河段的航道，改善航行条件。1927年，密西西比河下游发生的一场灾难性洪水造成200多人死亡，60多万人流离失所，此后，一项重大的河流工程项目开始启动。密西西比

图15 为了稳定发挥密西西比河重要交通线的作用，人类对其施加了诸多影响。图中的驳船位于路易斯安那州的巴顿鲁日附近，该处河道宽度大约700米

105

河干支流工程的建设目的是防洪和改善航运条件,河道裁弯取直是建设手段之一。河道人工裁弯取直使得河流长度缩短,河道坡度增大,流速增加。这样,水流侵蚀能力增强,河道变深,从而提高了河道的行洪能力。密西西比河长度的急剧缩短反映出工程作业的巨大规模。1929年,孟菲斯、田纳西和路易斯安那州红河码头之间的船只航行里程数为885千米,但到了1942年,由于一系列的裁弯取直,这一里程数减少了274千米,缩短了约30%。

密西西比河及其一些主要支流沿线近3 500千米的堤坝和防洪墙进一步提高了防洪能力,尽管做出了巨大努力,密西西比河仍然容易发生洪水。1993年,密西西比河上游发生的洪水是美国有史以来最为严重的自然灾害之一,4万多座建筑物被摧毁或遭受严重破坏。暴雨导致河堤决口1 000多处,由于河堤阻挡,洪水在洪峰过后无法返回河道,许多地方长时间受淹。同样很有可能的是,密西西比河洪水风险管理措施是墨西哥湾沿岸热带风暴破坏风险增加的原因之一。沿河修建河堤是海岸湿地减少的原因之一,泥沙和淡水的急剧丧失降低了湿地对风暴潮水位的抑制作用。这可能是2005年袭击新奥尔良市的卡特里娜飓风破坏加剧的原因。

城市河流

几千年前,世界上的一些大河的洪泛平原上出现了最初的城市文明(见第三章),从那时起,城市就对河流产生了很多影响。在印度河流域的哈拉帕和摩亨佐·达罗,考古发掘发现了陶瓷供水管道和位于街道下方的砖砌排水管道,人们认为,这些管道早在5 000年前就已经被投入使用。罗马人也以其先进的供水系统闻名于世。他们利用九条大型输水渠将远方的溪水和泉水输送到古罗马。其中一些输水渠长度超过60千米,需要在施工困难的山坡上开挖隧道以及用于检查和疏通清理的竖井。

这些早期市政系统的设计输水量很大,但重力的作用最终限制了实际输水量。水只能沿下坡方向从一处输送到另一处。现代文明利用能源抽水装置极大地提高了输水能力。例如,在美国西南部,科罗拉多河的河水经泵站抽送后的运输距离近500千米,途中穿过莫哈维沙漠后被输送到洛杉矶和圣迭戈等加利福尼亚西海岸大城市。

在城市化进程中,城镇地区的发展增长常常伴随着这种有心或无意为之的水系变化。人类可以对河流进行大规模的人工控制。例如,在日本,为防止新兴城市东京被淹没而实施的利根川东引工程将利根川向东改道了100多千米,17世纪中期这一历时50余年的宏大工程竣工,之后东京开始迅速发展。城市发展的早期阶段通常会对河流产生许多其他更微妙的影响。施工前清除树木和其他植被会导致降雨截留和蒸腾量减少,进而引起径流量增加以及裸露地表侵蚀加剧,常常造成河道内的沉积。107科学家对一些建筑工地的土壤侵蚀进行了监测,结果表明,其产沙量比自然条件下高出100倍。一个极端案例是马来西亚吉隆坡的一处废弃建筑工地,其年土壤侵蚀量超过60万吨,约为自然侵蚀率的2万倍。

城市发展早期阶段的另外一些影响源于人口的不断增长,这导致河流直接取水量或打井取水量增加,地下水位降低间接影响了河流水文情势。河流还为人类提供了现成的现代建筑材料,但河道采沙采石对河流的几何特征和生态具有重大影响。

城市对河流最重要的影响之一是城市化对洪水径流的影响。城市的大片区域被混凝土、石头、柏油和沥青覆盖,通常是不透水的。这往往会增加城市区域的径流量,而雨污管网又加剧了这一影响。由于雨水挟带的泥沙相对较少(同样是因为地表被不透水材料覆盖),所以进入河道后通常会产生侵蚀并拓宽河道。城市地区径流增加的另一个后果是大洪水发生频率增加。

河水污染一直是大型城市地区面临的一个问题,随着工业革命期间城市的发展,大量生活污水和工业废水排入河流,导致了特别严重的水污染问题。19世纪上半叶,随着城市人口的增长,抽水马桶数量迅速增加,伦敦泰晤士河水质不断下降。未经处理的污水直接排入河流,连同泰晤士河沿岸越来越多的工厂、屠宰场、制革厂和其他工业部门排放的废液。

农产品加工业排出的污水和废水等有机废液在有氧条件下可以被细菌和其他微生物分解。过量的有机废液会导致河流溶解氧含量降低,危害鱼类和水生植物并可能导致其死亡。到1849年,包括整个伦敦河段在内的泰晤士河感潮河段中的鱼类全部消失。当时,公共饮用水的水源依然是河水,和水有关的疾病横行:1830—1871年间,伦敦暴发了五次霍乱流行。1858年的夏天漫长而干燥,这一年史称"大恶臭年",由于河水散发出可怕的恶臭,国会大厦不得不休会数日。

对国家政治活动的这种直接影响也产生了一些积极作用,一些污水处理厂开始建设,到了1890年代,泰晤士河的水质已经有所改善。然而,在20世纪上半叶,污水处理和储存量跟不上伦敦人口增长,伦敦桥下游20千米处的河水含氧量在许多夏季都为零。1950年后，随着污水控制更加严格，处理设施得到改善，水质逐渐转好。到了1970年代，泰晤士河的水质得到广泛认可。1974年，人们从泰晤士河中捕获了1833年以来的第一条鲑鱼，各家媒体竞相报道。

许多工业化国家中流经主要城市的河流也有类似经历：伴随着工业化和人口增长，污染迅速增加，然后及时实施污染控制措施，环境质量恢复到可接受的程度。21世纪早期，一些污染最严重的城市河流位于亚洲的快速工业化国家。这些河流包括孟加拉国达卡的布里甘加河、菲律宾大马尼拉的马里洛河、印度尼西亚雅加达附近的芝塔龙河，以及流经中国众多城市的长江。

控制河盲症

洪涝是人类社会面临的与河流相关的最普遍的灾害，但在世界某些地区，一种被称为盘尾丝虫病或河盲症的疾病则是一个持续时间更长的问题。这种疾病由一种寄生虫引起，其传播者是在湍急的河流和溪流中滋生的小黑蝇。蠕虫一旦进入人体，就会在皮肤上形成变形的结节，而它的幼虫会移动，到达眼部后就会导致失明。据世界卫生组织估计，全世界有超过1 700万人受到感染，其中约50万人视力受损。

河盲症发生在热带非洲、拉丁美洲和阿拉伯半岛的部分地区。几乎可以肯定的是，拉丁美洲出现这种寄生虫是感染者移居美洲的结果，奴隶贩卖可能是原因之一。尽管1970年代初启动的一项大型河盲症控制方案取得了巨大成功，但西非依然出现了流行程度最高、临床表现最为严重的感染。西非盘尾丝虫病控制计划的重点是向西非的大片河流喷洒杀虫剂，杀灭黑蝇，控制该疾病传播。计划的高峰段涉及11个国家100多万平方千米区域内的5万多千米长的河流。喷洒频率很高，几乎每周一次，每年进行10～12个月，有些区域持续喷洒了20年以上，其目的是在这种人体寄生虫的寿命期内（一般认为10年以上）阻止其传播。

人们认为，这一极具雄心的计划使得西非约4 000万人免遭河盲症之苦，同时在曾经感染的河谷中开辟了25万平方千米可用于重新安置和种植的土地。监测表明，经消杀的河流中的其他昆虫和鱼类几乎没有受到有害影响，河流生态学家目前的看法是，这些河流中的其他生物受到永久性损害的可能性不大。

全球变暖

人类引起的全球气候变暖开启了社会影响河流的新纪元。温度总体上升将融化冰雪，植物的蒸发和蒸腾作用加大导致土壤水分损失增加。河流流量也会受到降水量、雨强和历时、降雨时间以及降水类型诸因素的变化的影响。气候学家认为，进入21世纪后，世界各地极端天气事件（包括热带气旋、干旱、热浪和暴雨）发生的频次、范围和强度可能进一步增大。所有这些都将不可避免地导致河流的变化。植物群落响应气候变暖的方式也会导致不那么直接但可能同样重要的变化。可以预料的是，为了应对气候变化的其他方面，社会也将会对某些河流施加更大的影响，例如，扩大干旱多发区的灌溉系统。

厘清全球变暖对河流的影响绝非易事，因为区分气候变化影响与各种河流特征的自然变化难度很大，同时还需要考虑土地利用和其他人类活动等可能的其他变化原因。尽管如此，全球变暖的影响在若干水系的某些近期变化中已经得到确认。对世界上很多大型流域的研究表明,20世纪发生特大洪水(重现期为100年)的风险显著上升。气温升高也对世界上许多地方的冰川产生了可预测的影响,即融化和退缩。喜马拉雅山和西藏部分地区的冰川目前正在飞快消退,由于印度、孟加拉国、尼泊尔和中国的河流依赖冰川融水补给,因此这些地区对数亿人口的远期供水问题产生了担忧。

自19世纪中期以来,北美洲和欧亚大陆大部分河流的冰盖普遍衰退,因为逐渐变暖意味着封冻日期推迟,解冻日期提前。以俄罗斯顿河下游为例,在大约100年的时间里,冰期长度缩短了整整一个月。芬兰托尔尼奥河的观测资料可以追溯到1692年,资料显示,整个观测期内解冻日期呈现提前的长期趋势。然而,这种趋势不具普遍性。西伯利亚中部和东部的河流呈现明显的相反趋势,即封冻日期提前,解冻日期推迟,封冻期延长。

由于无冰期延长以及降水量增加,流入北冰洋的北半球河流的输水量随之增加。北极地区淡水增加可能减缓或关闭所谓的"热盐环流",热盐环流是一条将大量暖水输送到北大西洋地区的洋流传送带。该环流由海水密度的差异触发,受温度和盐度控制,因此淡水增加能够减缓其流动。热盐环流有助于调节北欧气候,使气温高于纬度预期温度。

相反,自20世纪中期以来,其他很多河流每年的输水量都在下降。一些为大量人口提供水源的主要河流的流量不断减少,引发了人们对未来供水的进一步担忧。这些河流包括中国北方的

黄河、印度的恒河、西非的尼日尔河和北美洲的科罗拉多河。

干旱被认为是全球变暖引起亚马孙流域变化的最大推手。

关于该地区未来气候的很多计算机模拟模型表明，旱季降雨量将减少，而气温上升将进一步加剧降雨减少所带来的影响。干旱概率增加将对森林生态系统和流经的河流产生各种连带影响，包括发生火灾的可能性增大。可以预料，当地居民、野生动物和河流本身将面临严重的后果。

在欧洲，有人预测全球变暖将导致莱茵河的流量呈现更强的季节性。计算机模型估算表明，到2050年，莱茵河夏季平均流量减少可达45%，冬季平均流量增加可达30%。莱茵河夏季月份水量减少，主要与预测降水减少和预测蒸散发量增加有关。降水增加、积雪减少和融化提前使得冬季流量增加。因此，莱茵河冬季洪水造成的危害肯定会增加。河流季节性增强也将对莱茵河的生态产生很多影响。

河流恢复

人类活动在诸多方面有目的地或间接地影响了河流，为了弥补这些影响，许多国家也在努力进行"河流恢复"，扭转人类活动的一些早期影响。改善河流条件的尝试本身并不新鲜，本章前面提到的清理伦敦泰晤士河就是例证，但在20世纪末、21世纪初，各国广泛采取恢复、修复和缓解措施，这被认为是河流管理的一个独特阶段。恢复工程一般包括对河流毁损的修复，通常是为了更好地满足社会对自然、生态健康水道的需求和期望。

然而，让河流恢复到"自然"或"原有"条件通常困难重重。但至少在理论上，人们可以基于受人类影响之前河流沿线的历史条件，或者情况类似但受人类影响较小的参考河流沿线的条件来开展这一工作。然而在实践中，条件适当的参考河流可能

113

并不存在，或者历史基准时段以后的流域条件（如气候或植被）可能发生了变化。诚然，河流在各种自然条件下都会发生变化，而确定哪些变化是自然的，哪些是人类施加的，并不总那么简单。此外，虽然可以确定哪些人类影响是不利的，但要完全阻止它们可能更为复杂。

上述以及其他种种限制意味着重新恢复人类定居之前可能存在的条件几乎是不可能的。更合适的做法是恢复那些能够自我维持并与周围景观融为一体，因而通常更接近于自然状态的河流。因此，例如，法国卢瓦尔河大自然计划，世界上规模最大的河流恢复项目之一，其目的是确保在示范点保护典型的卢瓦尔河谷生态系统（包括泥炭地、峡谷、冲积平原森林和牛轭湖），并保持其生态功能。恢复河狸和鲑鱼等标志性河流物种是该项目的重要组成部分。

即使河流恢复项目的目标很明确，在大多数情况下，它们仍然需要与对河流的其他需求相平衡。其中一些要求可能是相互矛盾的。例如，一些环保主义者认为，河流整治和环境保护在本质上是不相容的，因为整治改变了原始野生动物群落赖以生存的自然环境。事实上，在某些情况下，生物的生态需求被破坏或改变的程度超过了它们适应的极限，生物无法生存。河流管理与其他自然环境管理问题一样，也需要妥协。当今世界，人口和经济增长似乎不可避免，更不用说人类诱发的气候变化包罗万象的影响，这些妥协因此可能会变得越来越微妙。

后　记

　　实际上，本书的每一位读者都会与某条河流存在某种联系，或者还不止一条河流。这种联系可能是生活在洪泛区，也可能是作为垂钓者或通过某种管道系统直接受益于河流的流动。河流影响着人类社会的方方面面，地球上很少有不受河流影响的地方，无论这种影响是直接的还是间接的，是当下的还是历史的。

　　赞美河流的多样性始终是本书的目标。河流丰饶又反复无常，其象征意义也因人而异，有时是矛盾的，有时是互补的。河流构成了无数生态系统的重要组成部分，滋养着城市和乡村。这种滋养既是精神上的，也是物质上的。人们对河流既崇拜又敬畏，既尊重又畏惧。从汹涌的洪流到潺潺的小溪，河水为艺术家、科学家、哲学家和将军提供了思想动力。从很现实的意义来说，人类历史很大的一部分都在河岸上展开。

　　古希腊哲学家以弗所的赫拉克利特断言："人不能两次踏入同一条河流。"所有河流本质上都是动态的。蜿蜒型河道可能突然变成辫状河流，涓涓细流可能变成汹涌波涛，冲出河岸淹没平原。从高山之巅到泥泞的三角洲，从蜉蝣幼虫的短暂一生到长江白鱀豚走向灭绝的漫长历程，无论是空间上还是时间上，河流所维持的生命无不处于动态变化之中。从最早的河船制造到工业污水充斥水道，人类对河流的利用和滥用也同样具有多样性和动态性。

河道只占地表面积的很小一部分，但其影响与这一直接的足迹却完全不成比例。无论怎样看待河流，人们都会承认河流主题的广泛性和多样性。同时，河流也反映了我们这个星球的
自然历史和社会历史。

河流

索 引

（条目后的数字为原书页码，
见本书边码）

D

E

F

河流

河
流

索引

索引

Nick Middleton

RIVERS

A Very Short Introduction

This book is for Cherry

Contents

List of illustrations

Introduction

Rivers flow on every continent and on all but the smallest island. They occur with an almost bewildering variety, ranging from a mere trickle to a mighty surge. As a source of water, rivers have always been objects of wonder and practical concern for people everywhere. They have acted as cradles for civilization and agents of disaster. A river may be a barrier or a highway. It can bear trade and sediment; culture and conflict. A river may inspire or it may terrify.

This book shines a light on the numerous roles that rivers have played in the life of our planet and its inhabitants, highlighting their importance to facets both obvious and obscure, ranging from sanitation to ichthyology, via divinity and literary criticism. The flow of rivers has inspired poets and painters, philosophers and scientists, explorers and pilgrims. No one can hope to understand the city of London without an appreciation of the River Thames, nor Egypt without the Nile. Rivers have lent their names to countries and determined the outcomes of wars.

A river can cleave a deep canyon and twist like a giant snake across its plains; plunge over great cliffs and stretch fingers of earth into the oceans. Rivers dominate landscapes, eroding and creating them. They are, without doubt, the product of a complex suite of natural processes. But the evolution of many rivers has been driven as much by social systems as by natural ones, surprising though this may at first seem.

Physically, people have long interacted with rivers, extracting their water and fish, modifying them to suit their needs. Rivers, in turn, have influenced innumerable aspects of culture through the ages, generating both myths and hydro-energy. Rivers have their place in legend, religion, and many other aspects of society, including music, art, and poetry. They are, therefore, not simply physical objects, part of the material world, but also cultural entities which interact with the social system. In many ways, rivers convey values as much as water.

It may not be surprising, then, to learn that precisely defining a river is not an easy task. Our friend the *Oxford English Dictionary* has it that a river is 'a copious natural stream of water flowing in a channel to the sea or a lake etc'. This definition serves for many rivers but not for all. Rivers in very cold places do not flow all the time. Neither do most rivers in deserts. In the former case, the water is frozen for lengthy periods; in the latter, there is often no water at all. The word 'copious' is tricky, too. Many readers will make a distinction in their minds between a river and a smaller body of water, such as a brook or stream. However, not all do. In legal terms, the word 'river' usually includes all natural streams, no matter how small. More difficult still is the word 'natural'. People have been interacting with rivers and changing them for thousands of years, so that today there are not many that can be described as completely natural and unmodified. To complete our deconstruction of the dictionary definition, let us remember that not all rivers flow into the sea or a lake. Some rivers disappear into the ground or dry up completely before reaching another body of water.

So, contrary to what we expect of a dictionary, its definition is not universal. This is not a modern conundrum. Dr James Clyde, whose book *Elementary Geography* reached its 25th edition in the late 19th century, was presented with it in his article on 'Rivers

1. Rivers play a primary role in shaping landscapes, sometimes in dramatic ways, as here in the Himalayan mountains in Nepal

and Rivers' in the *Scottish Geographical Journal* of 1885. He ducked the issue by paraphrasing John Stuart Mill in renouncing the metaphysical nicety of definition: 'every one has a notion, sufficiently correct for common purposes, of what is meant by river'.

There is a body of geographical research into how small children perceive their surroundings, and their notions of what a river should be are as good as any. Some of the best taken from a recent study are: 'wet water running down', 'a long blue thing that's wet', and 'something that flows and has fish and water'. All are sufficiently correct for our purposes.

Chapter 1
Nature's driver

Water is the driver of Nature.

> Leonardo da Vinci (1452–1519)
> (Italian painter, architect, and engineer)

We live on a wet planet. Water is the most abundant substance on Earth and covers two-thirds of its surface. It is also found in smaller quantities in the air we breathe, the plants and animals we see, and the ground on which we tread. This water is continuously on the move, being recycled between the land, oceans, and atmosphere: an eternal succession known as the hydrological cycle. Rivers play a key role in the hydrological cycle, draining water from the land and moving it ultimately to the sea.

Any rain or melted snow that doesn't evaporate or seep into the earth flows downhill over the land surface under the influence of gravity. This flow is channelled by small irregularities in the topography into rivulets that merge to become gullies that feed into larger channels. The flow of rivers is augmented with water flowing through the soil and from underground stores, but a river is more than simply water flowing to the sea. A river also carries rocks and other sediments, dissolved minerals, plants, and animals, both dead and alive. In doing so, rivers transport large amounts of material and provide habitats for a great variety of wildlife. They carve valleys and deposit plains, being largely responsible for shaping the Earth's continental landscapes.

Rivers change progressively over their course from headwaters to mouth, from steep streams that are narrow and turbulent to wider, deeper, often meandering channels. From upstream to downstream, a continuum of change occurs: the volume of water flowing usually increases and coarse sediments grade into finer material. In its upper reaches, a river erodes its bed and banks, but this removal of earth, pebbles, and sometimes boulders gives way to the deposition of material in lower reaches. In tune with these variations in the physical characteristics of the river, changes can also be seen in the types of creatures and plants that make the river their home.

Its narrow, linear form and its flow in just one direction provide an obvious spatial dimension to how we should describe and understand the physical, chemical, and biological properties of a river: horizontally from upstream to downstream. But a river is not just a channel. It is also an integral part of the countryside through which it flows, so a lateral dimension, to the surrounding landscape, is also appropriate. The links with the landscape, or riverscape as some would prefer, are innumerable. They range from the simple fact that most water in a river arrives in the channel after flowing across the surrounding topography to the importance of salmon in a river, say, as a seasonal source of food for local bears.

A third dimension is vertical. Rivers interact with the sediments beneath the channel and with the air above. The water flowing in many rivers comes both directly from the air as rainfall – or another form of precipitation – and also from groundwater sources held in rocks and gravels beneath, both being flows of water through the hydrological cycle.

The vital fourth dimension, time, also has an important place in river research. This is because of profound variations in many factors that affect rivers, not least the amount of water flowing in them. This varies on a wide range of timescales, from an intense rainstorm that lasts less than an hour to the effects of tectonic forces that operate over many millions of years.

Rivers are found all over the world and have left their mark on virtually every landscape. Certain areas lack surface drainage, but in some of these regions rivers flow beneath the land surface. In deserts, many rivers remain dry for most of the year, only channelling water in response to a sporadic rainstorm. Elsewhere, fossil channels and valleys indicate where rivers have flowed at some time in the more distant past. Such fossilized features also occur on other planets: channels and valleys have been identified on Mars and on Titan – the largest of Saturn's moons – and these networks are remarkably similar to river and stream features on Earth. On the surface of Mars, these features have been sculpted by flowing water in times past, but the river channels and drainage networks on Titan are thought to have been formed by the flow of liquid methane. For most of our planet's land surface, a flowing river of water is one of the most fundamental elements. Supplied with energy from sunlight and gravity, it is a feature that moulds valleys and slopes and provides a complex habitat for living communities.

River hierarchies

One interesting aspect of rivers is that they seem to be organized hierarchically. When viewed from an aircraft or on a map, rivers form distinct networks like the branches of a tree. Small tributary channels join together to form larger channels which in turn merge to form still larger rivers. This progressive increase in river size is often described using a numerical ordering scheme in which the smallest stream is called first order, the union of two first-order channels produces a second-order river, the union of two second-order channels produces a third-order river, and so on. Stream order only increases when two channels of the same rank merge. Very large rivers, such as the Nile and Mississippi, are tenth-order rivers; the Amazon twelfth order.

Each river drains an area of land that is proportional to its size. This area is known by several different terms: drainage basin, river basin, or catchment ('watershed' is also used in American English, but this word means the drainage divide between two adjacent basins in British English). In the same way that a river network is made up of a hierarchy of low-order rivers nested within higher-order rivers, their drainage basins also fit together to form a nested hierarchy. In other words, smaller units are repeating elements nested within larger units. All of these units are linked by flows of water, sediment, and energy.

Recognizing rivers as being made up of a series of units that are arranged hierarchically provides a potent framework in which to study the patterns and processes associated with rivers. At the largest scale, the entire river basin can be studied. Within the basin, at progressively smaller scales, a researcher may focus on a particular segment of a river between tributaries, a reach within a segment, and so on all the way down to a small patch of sand grains on the river bed. This hierarchical approach also emphasizes that processes operating at the upper levels of the hierarchy exert considerable influence over features lower down in the hierarchy, but not the other way around. At the river basin scale, important factors are climate, geology, vegetation, and topography. These factors have an influence at all lesser scales, down to the small patch of sand grains. That patch of sand also comes under other local influences, such as ripples in the flowing water, but these small variations in the current have a negligible impact on the drainage basin as a whole.

There is an appropriate timescale associated with related spatial scales and these too can be arranged into a hierarchy. Generally, the larger the spatial scale, the slower the processes and rates of change. Changes in climate and geology, for instance, occur on lengthy timescales, such as hundreds to millions of years. The ripples in the water operate on much shorter timescales: milliseconds to seconds.

It is also important to remember that, in general terms, as size increases, so too does the complexity of factors influencing the landscape and the river running through it. Hence, the small catchment of a first-order river channel, for example, may well occur on one rock type and lie within one climatic region. A larger catchment is more likely to span several rock types and climatic regions and is therefore more complex.

Types of river

The numerical ordering scheme detailed in the previous section is one of many attempts to classify rivers. There is an enormous variety of different types of river, or 'fluvial' system (from the Latin word *fluvius*, a river), when we extend our area of interest beyond the river channel to include the entire drainage basin. Each river classification depends on the perspective of the investigator and hence the aspect of greatest significance. A biologist may focus on the distribution of particular groups of organisms such as fish or aquatic plants. Different species may be associated with different types of topography and geology, for example, hence rivers may be placed in categories such as 'mountainous', 'upland', 'lowland chalk', 'lowland sandstone', and 'lowland and upland clay'. Others have used selected chemical factors as a basis for classification. An example is pH, so rivers might be classed as being strongly acid, slightly acid, or alkaline. An authority concerned with nature conservation might combine all of these perspectives and more. A classification of rivers in England, Wales, and Scotland based on vegetation communities devised by the Nature Conservancy Council recognizes four main groups of rivers, ten types, and thirty-eight sub-types.

Another simple way of categorizing rivers is by size. Some authorities prefer the word 'stream' when referring to rivers at one end of the size spectrum. A large or big river (both words are in common use to signify the other end of the spectrum) is usually one with either a large drainage basin, a long course, one that transports a large volume of sediment, or has a great volume of water flowing in it. We have noted that there is a consistent relationship between river length and drainage basin area, although not between the other variables due to variations in basin geology, relief, and hydrology. Most people when asked to list the world's largest rivers would come up with a similar list for their top 10 or 20, but a perfect definition remains elusive.

The pattern formed in a landscape by a network of rivers is a familiar way of distinguishing between different types of river system. There are several common variations on the essentially treelike pattern of a drainage network, and various descriptive terms are used, including dendritic, radial, trellis, parallel, and rectangular. The primary influence on these patterns is the geology of the landscape.

2. This satellite image of the Central Siberian Plateau illustrates a typical drainage network. Snow at higher altitudes contrasts with the snow-free valleys, helping to accentuate the drainage pattern

An obvious way of categorizing different types of river is by their types of flow. A river channel that carries water at all times throughout the year is described as 'perennial', but this does not describe all rivers by any means. Some channels have water flowing in them only in particular seasons. These seasonal, or 'intermittent', rivers may be in regions with a severe winter in which river water completely freezes, or in regions with a distinct wet season. A river with an even less permanent flow of water is described as 'ephemeral' and consists of channels that flow only for hours or days following individual rainstorms. Rivers that arise and flow in deserts are typically ephemeral rivers. A fourth category is the 'interrupted' river, one that has permanent flow over short reaches throughout the year while most of the river is dry. While these distinctions are undoubtedly real, like most classification schemes in the natural world, the boundaries between different classes are better viewed as points on a continuum of flow regime types. This is because, for example, during an extended wet period lasting several years, an ephemeral river may exhibit the characteristics of a seasonal river, while during dry periods, the wet season flow of a seasonal river may be absent or more intermittent, making it appear more like an ephemeral river.

How long is a river?

Measuring the length of a river is more complicated than it sounds. Measurements and estimates for the length of the world's rivers vary greatly depending on all sorts of factors, including the season of the year, the abilities of the cartographer, and the quality of his equipment, as well as decisions about what exactly is measured. In theory, the exercise should be straightforward: determine the position of the source, identify the mouth, and accurately measure the length of the river between the two. Finding the mouth is usually clear-cut. Its exact location is commonly defined as the intersection between the central line of the river and a line drawn between the two sides of the outlet.

Determining the exact location of the source is often more difficult. Searches for the source of particular rivers in remote and inaccessible regions have intrigued and inspired explorers for centuries, and continue to do so even today.

Disagreements about the true source of many rivers have been a continuous feature of this history of exploration. In one sense, a mission to find 'the' source of a river is destined to be a matter of conjecture simply because most rivers typically have many tributaries and hence numerous sources. For most authorities, the source that is farthest away from the mouth is considered to be 'the' source of the river, thus giving a maximum river length. But, unsurprisingly, differences arise as to the farthest source.

Another complicating factor is whether or not to include tributaries that have been given different names. In practice, the series of decisions made about the inclusion or exclusion of tributaries is probably the major element of a quest to find the source of a river, and these decisions represent one of the central reasons why not all measurements for a particular river agree. Take the Mekong as an example. Everyone acknowledges that the river originates on the Tibetan Plateau, but where exactly is open to debate. Candidates for its source include glaciers on the Guozongmucha Mountain, Lasaigongma Mountain, Zhanarigen Mountain, Chajiarima Mountain, and Mount Jifu. Others include Rup-sa La Pass, Lungmo Pass, and Lake Zhaxiqiwa. Given the number of designated sources, perhaps it is not surprising that the Mekong is variously referred to as the ninth longest and the twelfth longest river in the world, and that is without going into similar confusion surrounding many of the other major world rivers. Respectable texts give the length of the Mekong as anything between 4,180 kilometres and 4,909 kilometres. If we accept that the river's source is on Mount Jifu, which many do not, the river has six names along its 4,909-kilometre length. On the flanks of Mount Jifu, melted snow and ice flow as a stream named the Guyong-Pudigao Creek (which only flows in the summer). After just over 20 kilometres, this becomes the Guoyong River, which becomes the Zhaa River. The Zhaa merges with the Zhana River to become the Zha River, which becomes the Lancang River until it reaches the Chinese border with Myanmar, where it is known as the Mekong all the way to its delta in southern Vietnam. In its delta, the river splits into several branches that flow into the South China Sea.

Some say this is the Mekong in its entirety, and that it is 4,909 kilometres long. Others agree that it is 4,909 kilometres long but say that strictly the river should be called the Mekong- Lancang-Zha-Zhaa-Guoyong-Guyong-Pudigao. Another group would prefer to deal only with the stretch that carries the name Mekong, in which case the river is actually just 2,711 kilometres long. Many others differ more profoundly because for them the source is not on Mount Jifu at all.

If you are bemused, it is understandable. But it gets more confusing. Some rivers do not have a mouth. The Okavango River in southern Africa gradually diminishes into the inland Okavango delta, the size of which varies with the seasons. Hence, the exact point where the river ends changes seasonally. Some rivers have more than one channel. The length of which channel should be measured in a 'braided' (see below) stretch of river? The timing of measurement is also important. Guyong-Pudigao Creek on Mount Jifu only has water flowing in it during the summer melt season. Should it be counted if the flow is not continuous? Another difficulty of timing occurs in rivers that flood seasonally. When large stretches of the Amazon, for example, flood in the wet season, water that flows round a meander in the dry season flows more directly 'overland'. Should the length of the meander be counted, or not? Over longer periods of time, rivers can create new land, by depositing sediment in deltas, for example, so increasing their length.

Yet another important part of measuring the length of a river is the scale at which it is measured. Fundamentally, the length of a river varies with the map scale because different amounts of detail are generalized at different scales. The terrain along the course of a river has great complexity, with details nested within details. This geometric complexity, a quality known as 'fractal' that is inherent in many natural things, can be taken to the absurd. But when does a desire for greater detail cross the boundary into the realms of the absurd?

The use of satellite mapping, along with Global Positioning Systems (GPS), to establish accurate source locations will continue to improve our ability to study river systems in their entirety, but subjective decisions about the scale of study and which tributaries to include and exclude will continue to mean that it is effectively impossible to say definitively which river holds the 'world's longest' title. The Amazon and the Nile have been the main contenders for centuries as knowledge has improved and conventions have changed. The Scottish explorer John Hanning Speke thought he had solved one of the great mysteries of 19th-century world geography when he claimed in 1858 to have discovered that the source of the Nile was Lake Victoria. For much of the 20th century, most authorities recognized the Nile as the world's longest river, having added the longest tributary leading into Lake Victoria from the south. However, since the 1990s several credible claims have been made for the Amazon to be longer, following a number of expeditions in search of its source in the mountains of southern Peru. These claims put the length of the Amazon at some 6,850 kilometres, at least 150 kilometres longer than the Nile, but the debate is unlikely to end there.

River flow

Two particularly important properties of river flow are velocity and discharge – the volume of water moving past a point over some interval of time, although confusingly this may also be called simply the flow. A continuous record of discharge plotted against time is called a hydrograph which, depending on the time frame

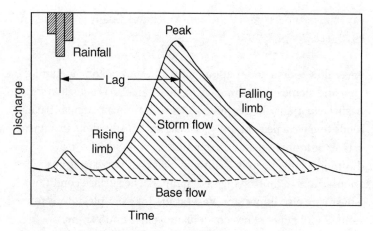

3. A storm hydrograph showing river discharge changing in response to a rainfall event

chosen, may give a detailed depiction of a flood event over a few days, or the discharge pattern over a year or more.

Measuring the flow of rivers and analysing the records is important for evaluating water resources and the assessment of flood and drought hazards. No river has a longer hydrological record than the Nile, where a water-level gauging structure was built on Roda Island at Cairo in AD 641. The official in charge of the Roda 'nilometer', the Sheikh el Mikyas, had the duty of observing the water level and during times of flood announcing the daily rise via public criers. This was always a tense time of year in Egypt. If the river did not reach a certain level, much cropland would go without water and famine could be expected, but at a certain higher level irrigation was assured, and with it taxes to the government. The position of Sheikh el Mikyas continued for more than 1,000 years. The last holder of the post died in 1947, and in the 1950s the Egyptian government decided to construct a major dam on the Nile at Aswan, thus significantly changing the country's intimate relationship with its river. The record from the Roda nilometer was invaluable in calculating the required storage capacity of the Aswan High Dam which was finally completed in 1970 (see Chapter 5).

River flow is dependent upon many different factors, including the area and shape of the drainage basin. If all else is equal, larger basins experience larger flows. A river draining a circular basin tends to have a peak in flow because water from all its tributaries arrives at more or less the same time as compared to a river draining a long, narrow basin in which water arrives from tributaries in a more staggered manner. The surface conditions in a basin are also important. Vegetation, for example, intercepts rainfall and hence slows down its movement into rivers.

Climate is a particularly significant determinant of river flow. It is the major factor controlling the different types of flow identified above: perennial, intermittent, ephemeral, and interrupted. All the rivers with the greatest flows are almost entirely located in the humid tropics, where rainfall is abundant throughout the year. These are the Amazon, the Congo, and the Orinoco, each of which discharges more than 1,000 cubic kilometres of water into the oceans in an average year.

Rivers in the humid tropics experience relatively constant flows throughout the year, but perennial rivers in more seasonal climates exhibit marked seasonality in flow. The Indus River receives most of its water from the Himalayan mountains and the maximum summer discharge is over 100 times the winter minimum due to the effect of snowmelt. Minimum discharge is often zero in rivers flowing largely in areas of high latitude and high altitude, where temperatures fall below freezing point for a portion of the year. In these intermittent rivers, the distinct contrast between minimum flow during the frozen winter and great floods during the summer melt season is regular and predictable.

By contrast, the flow of ephemeral rivers, typically found in desert areas, is spasmodic and unpredictable. This is because ephemeral rivers respond to rainfall which is notoriously difficult to predict in many deserts. One study of a river bed in the northern Negev Desert in Israel showed that on average the channel contained water for just 2% of the time, or about seven days a year. Some desert rivers can go for an entire year without any flow.

Year-to-year variations in river flow are also greatest in dry climates, whereas perennial rivers in the humid tropics have relatively steady flows from one year to the next. Records of discharge in the middle reaches of the Kuiseb River in the Namib Desert in Namibia over several decades show that flow has varied from 0 to 102 days per year.

Over longer periods, changes in rainfall and temperature have also resulted in changes in river flow regimes, although human interference has confused the picture in many cases (see Chapter 5). One of the clearest recent changes in natural flows is in West Africa where the desert-marginal belt to the south of the Sahara known as the Sahel experienced a marked desiccation of the climate over the last few decades of the 20th century, a trend that has continued into the 21st century. The flow of the Senegal River measured at Bakel, near the meeting of the borders between Senegal, Mauritania, and Mali, showed a marked decline towards the end of the last century. The average annual discharge at Bakel for the period 1904–92 was 716 cubic metres per second, but that average was just 379 cubic metres per second over the period 1972–92. The flow in 1984, a particularly dry year, averaged out at 212 cubic metres per second. A similar picture has been seen on the Niger River.

Some rivers are large enough to flow through more than one climate region. Some desert rivers, for instance, are perennial because they receive most of their flow from high rainfall areas outside the desert. These are known as 'exotic' rivers. The Nile is an example, as is the Murray in Australia. These rivers lose large amounts of water – by evaporation and infiltration into

13

soils – while flowing through the desert, but their volumes are such that they maintain their continuity and reach the sea. By contrast, many exotic desert rivers do not flow into the sea but deliver their water to interior basins. In southern Africa, water from the tropical highlands in Angola flows in the Okavango River to the Okavango Delta, the large wetland area in the Kalahari Desert in northern Botswana. In central Asia, water from the Parmir Mountains flows into the Aral Sea via two of central Asia's major exotic rivers: the Syr Darya and Amu Darya.

Some rivers are thought to be very old. Evidence from sediments deposited near the mouth of the Amazon suggests that the river has been flowing across South America for 11 million years. Over such great periods of time, all sorts of factors change, of course, and some rivers come and go. An example of a river that has disappeared is the Channel River which flowed westward in northwestern Europe some 20,000 years ago in the area now submerged beneath the English Channel separating Britain and France. This was the height of the last Ice Age, when sea levels all over the world were much lower than they are today because more water in the hydrological cycle was present as ice. Most of the British Isles and all of Scandinavia were covered in a thick ice sheet at this time, and the Channel River was fed by meltwater that flowed in the rivers of southern England, including the Thames and the Solent, that lay just beyond the permanent ice. Among the Channel River's other tributaries to the south were the Seine, Somme, Maas, Rhine, and Elbe.

Such ancient river channels are not solely of academic interest. The world's richest gold deposits, in the Witwatersrand district of South Africa, were laid down in river systems more than two billion years ago. Gold carried by these rivers was deposited in gravels where the velocity of the flowing water slowed. These gravels, known to geologists as Witwatersrand conglomerate, have produced nearly 50,000 tonnes, or 40%, of the gold ever mined, and probably still contain over one-third of the world's unmined gold reserves. Rivers have also played a key role in creating the valuable diamond deposits that stretch along the western coast of southern Africa. Diamonds have been eroded from deposits inland and carried to the coastline by the Vaal and Orange Rivers for 100 million years and more. This fluvial transport is also thought to be beneficial to the quality of the diamonds found in coastal sediments because the stones tend to break down during transport, increasing the concentration of higher-quality diamonds.

Erosion, transport, and deposition

An important measure of the way a river system moulds its landscape is the 'drainage density'. This is the sum of the channel length divided by the total area drained, which reflects the spacing of channels. Hence, drainage density expresses the degree to which a river dissects the landscape, effectively controlling the texture of relief. Numerous studies have shown that drainage density has a great range in different regions, depending on conditions of climate, vegetation, and geology particularly. The value tends to be high in arid regions of sparse vegetation, in temperate to tropical regions subjected to frequent heavy rains, and in areas underlain by rocks that are difficult for water to penetrate.

Rivers shape the Earth's continental landscapes in three main ways: by the erosion, transport, and deposition of sediments. These three processes have been used to recognize a simple three-part classification of individual rivers and river networks according to the dominant process in each of three areas: source, transfer, and depositional zones.

The first zone consists of the river's upper reaches, the area from which most of the water and sediment are derived. This is where most of the river's erosion occurs, and this eroded material is transported through the second zone to be deposited in the third zone. These three zones are idealized because some sediment is eroded, stored, and transported in each of them, but within each zone one process is dominant.

The changes in a river's slope that occur between its upper and lower reaches are reflected in a graphical measurement known as the 'long profile'. This is a section through the channel from its headwater to its mouth and is typically concave in shape because the headwaters are steep and slope decreases progressively in a downstream direction. This generally smooth, concave-upwards form is sometimes interrupted by outcrops of hard rocks that produce locally steeper slopes. Rapids form in these areas and the velocity of the river increases, promoting greater erosion, which over a long time period wears down the obstruction. In a place where relatively soft rocks are overlain by much more resistant rocks, a waterfall may occur. The world's highest waterfall, Angel Falls, or Kerepakupai Merú, cascades over a very hard sandstone rockface in Venezuela and is an awe-inspiring 979 metres in height.

All of the sediment carried by a river ultimately comes from the erosion of surrounding slopes and water flowing across and through the land surface, but the immediate supply comes from the bed and banks of the river channel. The flow of water carries this sediment in three ways: dissolved material – such as calcium, magnesium, and other minerals – moves in solution; small particles are carried in suspension; and larger particles are transported along the stream bed by rolling, sliding, or a bouncing movement known as 'saltation'. This material is deposited when circumstances change in some way, such as the slope of the river bed decreasing, so reducing the river's energy and ability to carry its load. Much of it is deposited in the sea. Globally, it is estimated that rivers transport around 15 billion tonnes of suspended material annually to the oceans, plus about another 4 billion tonnes of dissolved material.

In its upper reaches, a river might flow across bedrock but further downstream this is much less likely. Alluvial rivers are flanked by a floodplain, the channel cut into material that the river itself has transported and deposited. The floodplain is a relatively flat area which is periodically inundated during periods of high flow, typically every one or two years. When water spills out onto the floodplain, the velocity of flow decreases and sediment begins to settle, causing fresh deposits of alluvium on the floodplain.

Certain patterns of alluvial river channels have been seen on every continent and are divided at the most basic level into straight, meandering, and braided. Straight channels are rare in nature and, for the most part, are a function of the scale of assessment. They are described as straight at the regional scale, but at more local scales they are winding or sinuous to some degree. The most common river channel pattern is a series of bends known as meanders, named after the River Menderes in southwestern Turkey, which is well known for its sinuosity. Meanders develop because erosion becomes concentrated on the outside of a bend and deposition on the inside. As these linked processes continue, the meander bend can become more emphasized, and a particularly sinuous meander may eventually be cut off at its narrow neck, leaving an oxbow lake as evidence of its former course. Alluvial meanders migrate, both down and across their floodplain, a process that can be monitored by comparing old maps and repeated photography. This lateral migration is an important process in the formation of floodplains.

Braided rivers can be recognized by their numerous flows that split off and rejoin each other to give a braided appearance. These multiple intersecting flows are separated by small and often temporary islands of alluvium. Braided rivers typically carry abundant sediment and are found in areas with a fairly steep gradient, often near mountainous regions. The reason why one channel meanders and another is braided has been the subject of considerable research. Important factors that influence the channel pattern include the volume of water and velocity of flow, which are related in turn to the gradient of the channel and the

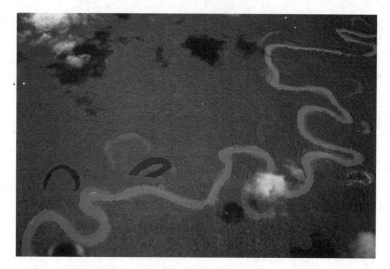

4. A meandering river and oxbow lakes in a remote part of New Guinea

nature of the channel, particularly the ease with which its bed and bank is eroded, which affects the supply of sediment to the river. These factors can change, over time and through space. For example, the Milk River in North America is a classic meandering river as it flows across southern Alberta in Canada but changes abruptly to a braided pattern shortly after entering Montana, USA. The change is probably due to differences in the material that makes up the bed and banks and a widening of the channel in the braided reach which reduces the power of the river.

The meander cut-off creating an oxbow lake is one way in which a channel makes an abrupt change of course, a characteristic of some alluvial rivers that is generally referred to as 'avulsion'. It is a natural process by which flow diverts out of an established channel into a new permanent course on the adjacent floodplain, a change in course that can present a major threat to human activities. Rapid, frequent, and often significant avulsions have typified many rivers on the Indo-Gangetic plains of South Asia. In

India, the Kosi River has migrated about 100 kilometres westward in the last 200 years, and the Gandak River has moved about 80 kilometres to the east over the last 5,000 years. The lower Indus River in Pakistan also has a history of major avulsions. Why a river suddenly avulses is not understood completely, but earthquakes play a part on the Indo-Gangetic plains.

Sometimes an avulsion can result in a channel being left dry, but on other occasions the channel becomes split, creating a river that flows in multiple channels. These multi-channel rivers are called 'anastomosing' or 'anabranching' rivers. At first sight, an anastomosing river can easily be confused with a braided river, which has a roughly comparable pattern. A braided river has multiple flows within a single channel, whereas an anastomosing river has multiple interconnected channels. Nonetheless, debates about the differences continue and numerous classifications of channel pattern are used. Misunderstandings can also arise when river flow is high or low. At high discharge, a braided river with submerged bars may look like a single-thread channel, and at low discharge an anastomosing river may carry water in a single main channel only, so appearing as a single-channel river.

Most rivers eventually flow into the sea or a lake, where they deposit sediment which builds up into a landform known as a delta. The name comes from the Greek letter delta, Δ, shaped like a triangle or fan, one of the classic shapes a delta can take. Examples of this type of delta include two of Africa's largest: those at the conclusion of the Niger and Nile Rivers.

The river provides the sediment that makes up a delta, but there are many other influences on its shape, including the volume of water flowing, the amount of sediment, and the relative importance of the flow of the river, the ebb and flow of tides, and the energy of waves. Fan-shaped deltas like the Niger and Nile are dominated by the action of the waves. Deltas dominated by the flow of the river typically extend further out into the sea as a lobe,

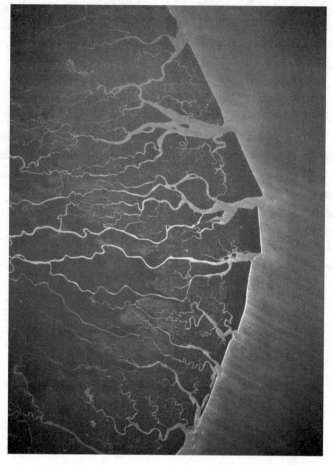

5. The archetypal fan-shaped delta of the Niger River, the largest delta in Africa

its channels branching like the toes or claws of a bird. The Mississippi River delta is an example of the 'bird-foot delta' type. Tide-dominated deltas form in locations with a large tidal range or fast tidal currents. They are typified by numerous islands elongated parallel to the main tidal flow and perpendicular to the shore line. Good examples are the deltas of the Fly River in New Guinea and the delta at the confluence of the Brahmaputra and Ganges Rivers.

Material laid down at the end of a river can continue underwater far beyond the delta as a deep-sea fan. The world's largest submarine fan lies beyond the Ganges-Brahmaputra delta. The Bengal Deep-Sea Fan is almost 3,000 kilometres long, more than 1,000 kilometres wide, and may be more than 16 kilometres thick at its deepest part. It is linked to the Ganges-Brahmaputra delta by a submarine canyon which funnels the sediment from the river to the deep-sea bed. The origins of the fan date from the collision of India with Eurasia, the tectonic event that created the Himalayan mountain range, making it more than 40 million years old.

River ecology

A great diversity of creatures makes up the ecology of rivers, an interconnected web of life that ranges from microscopic algae to huge fish larger than a human being. Their diverse communities reflect the great array of running water environments that vary from large lowland rivers occupying basins on a subcontinental scale to small, turbulent mountain brooks. The physical structure of the river is one set of influences on its ecology, but chemical and biological attributes are also important, and all are to some degree related. The water's oxygen content, acidity or alkalinity, nutrients, metals, and other constituents are all determined largely by the types of soil and rock that make up the drainage basin, but also in part by interactions with plants and animals both in the water and on land.

The organisms found in fluvial ecosystems are commonly classified according to the methods they use to gather food and feed. 'Shredders' are organisms that consume small sections of leaves; 'grazers' and 'scrapers' consume algae from the surfaces of objects such as stones and large plants; 'collectors' feed on fine organic matter produced by the breakdown of other once-living things; and 'predators' eat other living creatures. The relative importance of these groups of creatures typically changes as one moves from the headwaters of a river to stretches further downstream, reflecting physical factors such as channel width, shading by trees, and the velocity of the water. This is the 'river continuum concept' which describes a continuum of changes that integrate energy sources, food webs, and stream order in an essentially linear way. Hence, small headwater streams are often shaded by overhanging vegetation which limits sunlight and photosynthesis but contributes organic matter by leaf fall. Shredders and collectors typically dominate in these stretches, but further downstream, where the river is wider and thus receives more sunlight and less leaf fall, the situation is quite different. Here, food chains are typically based on living plant material rather than leaf fall, so there are few shredders and probably more predators.

The river continuum concept is a popular model that has influenced many studies of fluvial ecosystems, but it is not the only one. Another important model used to study river ecology stresses the importance of the annual pulse of floodwaters that extends many rivers in temperate and tropical regions on to their floodplains. The 'flood-pulse concept' broadens the focus beyond the main river channel and puts much greater emphasis on interactions with a greater variety of habitats such as the marshes and lakes typically found on floodplains. These habitats are broadly synonymous with the river's 'riparian' zone (from the Latin word *ripa*, a bank), made up of any land that adjoins, regularly influences, or is influenced by a body of water. Vegetation in the riparian zone helps to maintain the condition of aquatic ecosystems in several ways. These include providing bank stability and so minimizing erosion, filtering sediment, and processing nutrients from the drainage basin, particularly nitrogen. Fallen branches or trunks from riparian trees also create woody habitat areas for many fish and smaller creatures.

From the ecological perspective, the unidirectional flow of a river is a unique situation. Flowing water influences many aspects of the river environment, moving things and thus helping to disperse organisms and transport nutrients. Flowing water affects the shape of the channel and the nature of its bed, disturbing both on occasions of strong flow, maintaining a dynamic habitat for fluvial plants and animals. Rivers also deliver water, energy, sediment, and organic matter to marine ecosystems. This flow is overwhelmingly in one direction, but not entirely so. Some fish swim against the flow, migrating upstream to spawn, for instance. Fish that migrate from the sea into fresh water for breeding, so-called 'anadromous' species, such as salmon, are prime examples. Salmon attain most of their body mass feeding at sea, so when they die in a river after spawning their carcasses make an important contribution of nutrients and energy to both aquatic and adjacent terrestrial ecosystems.

The flow of water has almost inevitably produced an emphasis on spatial complexity in studies of river ecology, but variations in stream flow over time are also important. The quantity, timing, and variability of a river's flow create a mosaic of habitats to which fluvial organisms have adapted. The ecology of rivers in regions with a Mediterranean climate, for example, is attuned to substantial seasonal variability in flow because most of the rain falls in winter (often 80% or more in three months). A cool, wet season is followed by a warm, dry season which produces a rhythm of flooding and drying in the rivers, although the intensity of each season can vary markedly from year to year.

There's no doubting the numerous fundamental ways in which a river's biology is dependent upon its physical setting, particularly in terms of climate, geology, and topography. Nevertheless, these relationships also work in reverse. The biological components of rivers also act to shape the physical environment, particularly at more local scales. Beavers provide a good illustration of the ways in which the physical structure of rivers can be changed profoundly by large mammals. They cut wood and construct dams, trapping sediments and organic material, modifying nutrient cycles, and ultimately influencing many other communities of plants and animals.

Finally, it is worth emphasizing again the many ways in which the ecology of rivers has effects far beyond the channel itself. In the same way that a river plays a key role in shaping the landscape it moves through, its flow provides important services to many of the plants and animals that inhabit that terrain. The most obvious of these is as a source of water and sustenance. Flowing water both delivers and removes many vital nutrients and other constituents to and from ecosystems, but rivers also have effects that may be less immediately obvious. The distribution of many terrestrial plant and animal species concords with the geography of major river systems because rivers can act both as corridors for species dispersal but also as barriers to the dispersal of organisms. One of the first to recognize the importance of rivers as obstacles to the movement of certain creatures was the naturalist Alfred Russel Wallace, who in the mid-19th century defined distinct areas in South America bounded by major rivers in the Amazon Basin, each with its own distinct communities of species. This idea of the river acting as a barrier is one of a number of hypotheses put forward to explain the evolutionary origin of the astonishing richness of species found in Amazonian forests.

The Amazon: mightiest of them all

By almost every measure, the Amazon is the greatest of all the large rivers. Encompassing more than 7 million square kilometres, its drainage basin is the largest in the world and makes up 5% of the global land surface. The river accounts for nearly one-fifth of all the river water discharged into the oceans. The flow is so great that water from the Amazon can still be identified 125 miles out in the Atlantic: early sailors could drink fresh water from the ocean long before their first sighting of the South American continent. Nonetheless, the lower reaches of the Amazon flow down such a gentle gradient that the physical influence of sea tides can still be identified more than 1,000 kilometres upstream from the Atlantic.

The Amazon has some 1,100 tributaries, and 7 of these are more than 1,600 kilometres long. The main tributaries are often classified according to the colour of their waters, which also reflects their source. Black-water tributaries attain their tea colour from high levels of dissolved plant matter leached from low-lying areas of sandy soils. White-water rivers are coloured by the high loads of sediments transported from the Andes. The clear-water rivers carry low levels of sediments and organic matter from the crystalline rocks of the Guyana and Brazilian shields.

In the lowlands, most Amazonian rivers have extensive floodplains studded with thousands of shallow lakes. Up to one-quarter of the entire Amazon Basin is periodically flooded, and these lakes become progressively connected with each other as the water level rises. Researchers using GPS have measured a sizeable part of South America sinking by nearly 8 centimetres because of the extra weight due to flooding in the Amazon, an area that rises again as the waters recede. This annual rise and fall of the Earth's crust is the largest we have detected.

Many of the Amazon's plants and animals have adapted to living in an environment that is seasonally waterlogged, some areas for up to 11 months a year and to depths of 13 metres. Many tree species of the Amazon rainforest depend on the floods for seed dispersal, for example, either floating downriver or through fish species that are dependent on fruits and seeds. The great diversity of the Amazon's aquatic habitats has played a key role in producing the most diverse fish fauna on the planet. In total, with about 2,500 fish species that have been described by scientists (and probably more than 1,000 still awaiting description), the Amazon's species richness comfortably exceeds that of all other large river basins. Its two mightiest fish, the arapaima and the piraíba, each has a maximum weight of about 200 kilograms, more than twice that of an average man.

The one measure by which the Amazon is not generally regarded as the greatest of all rivers is length. It is comfortably the longest in the Americas, but most authorities place the Nile at the top of the world list. However, the difficulties in measuring the length of a river mean that debate on the matter will undoubtedly continue (see above).

The Onyx: an unusual river

The longest river in Antarctica, the River Onyx is just 32 kilometres in length and is in many respects quite different from rivers in most parts of the world. The Onyx is situated in the McMurdo Dry Valleys region, one of a small number of ice-free desert areas that occur along the coastline of an otherwise ice-covered continent. The climate is very dry and bitterly cold, with an average annual temperature of −20°C. The small amount of precipitation (less than 100 millimetres a year) that does fall comes only as dry snow and has virtually no direct effect on the river because fierce winds mean it never settles. Hence, the Onyx, and other rivers of the McMurdo Dry Valleys, flow for only four to ten weeks a year. This occurs during the summer when the temperature is warm enough to melt glacier ice, the only source of river water.

The Onyx flows from the Lower Wright Glacier into Lake Vanda, which has a salinity more than ten times that of seawater and a permanent cover of ice. There are no plants in the region and no fish or insects in the river, but worms, microscopic animals, and communities of algae growing as mats inhabit the river bed. These algal mats can survive long periods of desiccation, making the Onyx a relative hotspot of life in an otherwise barren landscape.

River floods

To hydrologists, the term 'flood' refers to a river's annual peak discharge period, whether the water inundates the surrounding landscape or not. In more common parlance, however, a flood is synonymous with the river overflowing its banks, and this is the meaning used here. Rivers flood in the normal course of events. This often occurs on the floodplain, as the name implies, but flooding can affect almost all of the length of a river.

Extreme weather, particularly heavy or prolonged rainfall, is the most frequent cause of flooding. The melting of snow and ice is another common cause. These events can often be predicted to an extent because they are seasonal. Other reasons for river floods are usually harder to anticipate. They include landslides, log jams, ice jams, avalanches, volcanic eruptions, and earthquakes.

River floods are one of the most common natural hazards affecting human society, frequently causing social disruption, material damage, and loss of life. The hazards associated with floods have encouraged the development of many techniques for predicting them. Flood hazard maps are commonly utilized for land-use zoning, enabling an authority to prohibit certain developments on land that is particularly flood-prone, for instance. Anticipating when a flood will occur can be done in several different ways. Most floods have a seasonal element in their occurrence and can often be forecast using meteorological observations, with the lag time to peak flow of a particular river in response to a rainstorm calculated using a flood hydrograph.

Other flood predictions seek to estimate the probable discharge which, on average, will be exceeded only once in any particular period, hence the use of such terms as '50-year flood' and '100-year flood'. It is a general rule that the magnitude of a flood is inversely related to its frequency, or probability, of occurrence (in other words, the larger the flood, the less likely it is). A flood that is likely to occur only once in a hundred years – the 100-year flood – has a 1% likelihood of occurring in any year, and the average interval between two floods of that magnitude is 100 years. For engineering purposes, it is useful to know the probability of a flood of a particular magnitude so that, for example, a bridge designed to last for 50 years can be built large enough to withstand a 50-year flood, and often a 100-year flood just in case. These are statistical probabilities, however, and there is still the chance that the bridge may be swept away by a far larger flood.

Many of the less predictable causes of flooding occur after a valley has been blocked by a natural dam as a result of a landslide, glacier, or lava flow. Natural dams may cause upstream flooding as the blocked river forms a lake and downstream flooding as a result of failure of the dam. Earthquakes, which can cause enormous landslides, are a particularly common cause of natural dams. For example, the Īnangahua earthquake in South Island, New Zealand, in May 1968 triggered a huge landslide that dammed the Buller River. The rising water backed up for 7 kilometres, raising the river 30 metres above its normal level. Fears that the dam might suffer a catastrophic breach led to an evacuation of all the people living in its path, but the river eventually overflowed the landslide dam, eroding it downward gradually without causing serious flooding downstream.

Breaches of natural dams account for most of the largest known floods of the last 2.6 million years, the so-called Quaternary Period. The biggest that we know about occurred as a result of ice-dam failure after pre-existing continental drainage systems were blocked by ice sheets during the Ice Ages that have characterized the Quaternary. Some of the largest ever to have occurred on Earth were the Missoula floods in the northwestern USA of today. They resulted from the repeated breaching of an ice dam that blocked the present-day Clark Fork River between about 18,000 and 13,000 years ago. The ice created an immense lake known as glacial Lake Missoula which spilled out to create the Missoula floods when the ice dam periodically failed. The peak discharge of the Missoula floods is thought to have been a gigantic 17 million cubic metres per second, more than ten times the combined flow of all the rivers of the world today.

The evidence for the Missoula floods is convincing, but it is one of several great floods known or suspected to have occurred in prehistoric and geological times and not all are as well substantiated. Somewhere at the confluence of fact and fiction lie hundreds of flood legends from cultures all across the world. These stories take their place among a far greater number of myths, sacred traditions, and beliefs based on the flow of rivers that form the subject of the next chapter.

Chapter 2
Sacred flows

Their Lord will guide them by their faith; there shall flow from
beneath them rivers in gardens of bliss.

The Koran 10.9

Throughout history, the flow of rivers has nurtured life and spread
fertility to countless societies while also, on occasion, bringing
death and devastation. This dual function, as a force of nature that
sustains life but also takes it away, has generated cultural echoes
in groups all over the world. The powerful hold that rivers have
over humankind has become embedded in innumerable
traditions, myths, and sacred rituals through the ages.

Mythical rivers

In Greek mythology, the land of the dead, or underworld, was
surrounded by five rivers. These were the Acheron (the river of
woe), Cocytus (the river of lamentation), Phlegethon (the river of
fire), Lethe (the river of forgetfulness), and Styx (the river of hate).
When somebody died, the spirit of the dead was ferried across the
water (in some cases the Acheron, in others the Styx) by the
boatman, on payment of a fee. Each new arrival in the underworld
was judged, determined to be good or bad, and transferred either
to a place of torment or to the Elysian Fields that can be equated
with Paradise. Residents of the Elysian Fields had the possibility

of rebirth once their previous life had been forgotten, a feat achieved by drinking the waters of the River Lethe.

Miraculous powers were associated with the River Styx. Its waters were used by the gods to seal unbreakable oaths, and the Greek hero Achilles was immersed in the river as a child, making him entirely invulnerable except for the spot on his heel where his mother held him for his dip. Achilles eventually lost his life when a poisoned arrow hit him in this heel, an episode that spawned the expression 'Achilles' heel', still used to describe a person's principal weakness.

Traversing a river to the underworld appears in other belief systems. The Sanzu River is the river to cross in the Japanese Buddhist tradition, and the Vaitarna River serves the same purpose in several Hindu religious texts, although only for sinners (those who do good deeds in life do not have to cross the river).

Rivers also feature prominently in accounts of Paradise, in Hebrew, Christian, and Islamic traditions. Early Christians adopted the Hebrew Bible and with it the story of Genesis, in which a single unnamed river is described as flowing out of Eden to water the garden, from where it emerges to feed four rivers. These rivers – the Tigris, Euphrates, Gihon, and Pishon – flow to different parts of the world. While the Tigris and Euphrates are known quantities, the Gihon and Pishon were a source of wonderment and confusion for many travellers in ancient and medieval times. The Pishon was long associated with Arabia and later became identified as the Ganges or Indus and, on occasion, the Danube. The source of the Gihon, by contrast, was commonly placed in Ethiopia and hence the river was equated with the Nile. The apparent impossibility of such widely disparate rivers as the Tigris, Euphrates, Ganges, and Nile all having a common source in the Garden of Eden was explained by the suggestion that these rivers flowed underground on leaving Eden initially, resurfacing at great distance from Paradise and from each other.

6. Map of the four rivers of Paradise, or Paradisus, published in a 16th-century book, *Chronicum, Scripturae autoritate constitutum*

Four rivers are also specified in the Paradise which the Koran says Allah has prepared for faithful Muslims, a place most frequently described as 'gardens under which rivers flow'. Here the devout are sustained by a river of flowing water, and three others comprising milk, wine, and honey. It is frequently thought that the four rivers of Paradise have exerted a great influence on the design of Islamic gardens, often created as earthly representations of Paradise. Many Islamic gardens are laid out in four sections, divided by channels of water fed from a pool or fountain at the garden's centre, but this four-part design with flowing water playing a defining role actually predates Islam. Hence, it is probably not the layout that reflects a specifically Muslim view of Paradise so much as the description of Paradise that reflects a pre-existing expression of garden form.

Rivers are an important feature in some of the earliest Sanskrit texts from India. One of the most prominent rivers mentioned in the Rig Veda, the first of four books that form the basis for the Hindu religion, is the Sarasvati, a river that is also personified as the goddess Saravati. As a river, the Sarasvati is described as large and fast-flowing in the Rig Veda, but later Hindu texts, including the Mahabharata, depict it as having been reduced to a series of saline lakes. Contemporary interest in the Sarasvati River has led several scholars to equate the mythical river with a number of ancient, dry river channels discovered with the aid of satellite imagery in India's Thar Desert in recent years.

Flood legends

Stories of a great flood crop up with uncanny frequency in the mythology of innumerable cultures, both ancient and modern. The deluge described in the Biblical book of Genesis is well known to many people in the Judaeo-Christian world and has numerous similarities with the flood described in the earlier Babylonian account of the Epic of Gilgamesh and similar stories from Sumeria and Assyria, also in Mesopotamia. The flood is explained as God's way of cleansing the Earth of wayward humanity, although one man and his family manage to escape in a boat, or ark, with representatives of the planet's wildlife population to keep them company. In all of the stories, the ark ends up on a mountain top and birds are sent forth to see whether the floodwaters have receded. The great flood has considerable symbolic significance, involving an obvious cleansing element as well as being a vehicle for rebirth, marking a clear break between the antediluvial and postdiluvial worlds. The event is effectively repeated at the personal level in various ceremonies of purification by water, including the Christian sacrament of baptism, in which the initiate is cleansed of the old ways in the waters of a river (or font) and is reborn in Christ. The ceremony mimics the baptism of Jesus in the River Jordan.

A similar flood-induced divide between a previous world and a new cosmological order appears in the written testimonies of several Maya groups from Central America. In a number of versions, the deluge occurs after a celestial caiman has been decapitated, interpreted as a flood caused by torrential rain. In some accounts, humanity continues thanks to a few survivors, but many other Mesoamerican flood myths, particularly those recorded by the Aztec peoples, tell of no flood survivors so that creation had to start again from the beginning.

A creation myth from Norse mythology tells how the world emerged at the meeting place of fire and ice, a great void into which eleven rivers flowed. An evil frost giant named Ymir sprang from this place and gave birth to the first man and woman from under his left armpit. Eventually, Ymir was killed by gods who created the world out of his body. His skull became the sky; his spilled blood became the Norse flood that drowned all of the frost giants with the exception of one man and his wife, who escaped in a vessel made of a hollowed tree trunk.

Floods also feature in myths and stories told by numerous Aboriginal groups in Australia, their prominence explicable at least in part by the often dramatic nature of flooding in desert landscapes. One story told by the Wiranggu of South Australia tells of a rain-maker named Djunban who was not fully concentrating on his rain-making ceremony one day and brought unusually heavy rain as a result. Djunban tried to warn his people, but a great flood came and washed them away with all their possessions, forming a hill of silt. This is the origin of gold and bones found in the hill.

The possibility that floods described in myths from all over the world are based on real events has on occasion engendered great debates. Deconstruction of the Biblical flood story, for example, played a central role in the rise of scientific geology in the 19th century. The British geologist Charles Lyell, in his influential book *The Principles of Geology* (published in three volumes, 1830–3), dismissed the prevailing belief in Noah's flood due to a lack of evidence in the geological record. Lyell's book was one of the key works in a struggle between science and faith as the predominant basis for explaining the origins of the world around us. It led to a widespread understanding that our planet is very much older than creationists believed.

Sacred rivers

Many belief systems have invested elements of the natural world with sacred characteristics, and specific rivers feature prominently among them. Rivers were sacred to the Celts of northwestern Europe, for instance, and many were personified as goddesses. Some of the river names used today in this part of the world can be traced back to the Celtic deities who lived near them or died in them. In Ireland, the Rivers Boyne and Shannon derive their names from goddesses who drowned in them after seeking wisdom from a magical well.

The importance of the Nile to the ancient Egyptians was reflected in a number of major and minor gods being associated with the river. Hapi was the god who personified the river's annual flood, the inundation of tears shed each year by the goddess Isis, in sorrow for her murdered husband. Hapi, the Nile deity responsible for collecting these tears, lived in a cataract near today's Aswan, surrounded by crocodiles and goddesses, some of whom were frogs, others women with frogs' heads. Each year, at the start of the flood, Egyptians carried out mass animal sacrifices to Hapi.

In numerous cases, the sanctity of a river is linked to a creation myth that arises from water's position as a primordial element. The River Birem in Ghana, for example, is considered to be the spiritual force and fountainhead of the Akyem kingdom because legend has it that the people of Akyem emerged from the depths of the river. Indeed, rivers, streams, and other water bodies all across Africa are frequently regarded as the habitat of deities and ancestors and hence treated with considerable reverence. The most prominent of the river divinities in Yoruba cosmology, for instance, is Yemoja, ruler of the Ogun River in Nigeria. Yemoja is the mother of all fish and the giver of children, and is customarily brought offerings of yams and chickens by women who want to start a family. In many parts of southern Africa, spirits who dwell in certain river pools are responsible for the creation of traditional healers (see below).

Many of the indigenous peoples of Siberia also traditionally enjoy close links to nature, in which rivers and other elements of the landscape are central to their animistic spiritual belief systems. Rivers, springs, lakes, and mountains are understood to have spirit-guardians whose presence must be regularly acknowledged and honoured via a community's shaman. For example, the Katun River is considered central to the culture of the indigenous Altaians who inhabit the Russian Altai region on the confluence of its borders with Kazakhstan, China, and Mongolia. Altaians consider the Katun to be a living being, and show appropriate respect in several ways. These include not throwing stones into the river, saying special words when crossing it, and not taking water from the Katun at night because this may upset the river's spirit.

A similar attitude toward rivers is found among the Mansi who live in the Tyumen region of northwestern Siberia. Sacred rivers such as the Yalbynya must not be fished, and even rowing a boat is prohibited in some stretches, so the vessel has to be pulled along from the bank. Other reaches come with different embargoes: the extraction of drinking water is forbidden, for example, or felling trees on certain banks. The river mouth is considered to be the most significant part of the Yalbynya, where local people throw money on passing.

In southeastern Europe, the waters of the Danube play an important part in traditional funeral customs practised by Bulgarians and Romanians living along the river's lower reaches. The river has considerable symbolic value for the idea of death as a long journey to the nether world and is incorporated into often elaborate memorial rituals. 'Freeing the water' of the deceased is a rite that provides water to the dead person for use in the afterlife. The ritual, which varies in detail from village to village, usually involves a child bringing river water to certain houses. In the Bulgarian village of Leskovec, the child is a girl who then returns to the Danube with several women where they lay down a tablecloth on the riverbank and set out a meal consisting of boiled wheat and wine. The women light a candle and hang gifts for the child on a forked stick taken from an apple tree. The girl puts her right foot in the river and asks three times for the ceremony to be witnessed, at which point a hollowed pumpkin containing a candle, some wheat, and a piece of bread is launched from the riverbank. When the pumpkin floats away down the Danube, the water will reach the deceased, but should the pumpkin turn over, the deceased will be angry.

Rivers feature among the most important types of sacred place in Hinduism. About 3,000 years ago, rivers were revered by the

37

Aryan people of the Vedic period in the region that is now India, and evidence from archaeological excavations suggests that the Hindu practice of mass bathing in rivers on auspicious occasions dates back to (and derives from) a similar practice in the Harappan civilization of the Indus Valley, up to 2,000 years before that. Indeed, the words 'Hindu' and 'India' are derived from the Indus.

Virtually all Indian rivers are revered as deities, but the Indus is commonly referred to as one of the seven holy rivers of India, the others being the Ganges, Yamuna (or Jumna), Sarasvati, Godavari, Narmada, and Kaveri. However, the Indus and the Kaveri are occasionally replaced by the Tapti and the Kistna. The rivers are often thought of as the veins in the earth's body, and many specific places along a river's course are particularly sacred, including the source, mouth, and confluences. The most sacred of all India's holy rivers is the Ganges.

The Ganges

The connections Hindus have with the Ganges provide one of the most striking examples of the sanctity of rivers. Indeed, in India 'Ganga' is both the name of the River Ganges and the personification of the river as a goddess. The holiness of the Ganges is enshrined in numerous Hindu epics and scriptures, including the Ramayana, the Mahabharata, the Vedas, and the Puranas. The story of the river's arrival on Earth from the heavens has it that the feat was achieved by a sage, known as Bhagiratha, who went to the Himalayan mountains and managed to persuade the river to descend. In several versions of the story, it is the god Shiva who controls the flow of the river, and Gangadhara, or 'Bearer of the Ganges', is one of Shiva's other names.

The water of the Ganges has numerous auspicious properties for Hindus. It acts as a medicine for every ailment, and bathing in it cleanses the devoted from all sin. Crucially, however, when a person's ashes or bones are entrusted to the river, the soul will be released for rebirth. For many Hindus, the holy city of Varanasi is the preferred place for this final transformation. The west bank of the Ganges at Varanasi is divided into many sections of river frontage each consisting of a series of long steps down to the water, the 'ghats' where people come to bathe, wash their clothes, and cremate the dead. About 80 corpses a day are burned at the two main ghats in Varanasi, most of these brought to the river from outside the city. The ashes of many more people are brought for final immersion in the Ganges. Some corpses that are not cremated, such as those who had smallpox in the past or who died of cholera, are simply weighted down and submerged in the holy waters. Among the most important ghats that specialize in cremation is Manikarnika, which contains the well dug at the beginning of time by Vishnu, one of the most significant Hindu deities who is sometimes depicted as a man-fish. This is the place where all creation, or the cosmos, will burn at the end of time.

Another of the Ganges' most sacred places occurs at its confluence with the Yamuna River, a pilgrimage site popularly known as Prayag, near today's city of Allahabad. This is one of four sites of the mass Hindu pilgrimage Kumbh Mela. According to legend, this is also the place where the mythical Saraswati River joins the Ganges and Yamuna Rivers, thus lending the confluence an additional level of sanctity. The full Kumbh Mela, in which many millions of devotees bathe in the Ganges to purify their sins, takes place every 12 years. The event in 2001 was thought to have been attended by some 60 million people, making it the world's largest gathering in recorded history.

Sacred river creatures

Given the reverence with which numerous rivers are viewed by peoples all over the world, it should come as no surprise to learn that some of the creatures found in rivers have also been the objects of respect and veneration. Certain types of fish found in

7. **Kumbh Mela, the mass Hindu pilgrimage to the sacred Ganges**

the Nile were surrounded by mythology and superstition in ancient Egypt, where some species were sacred in particular settlements but not in others. The mormyrus, recognizable by its lengthy down-turned snout, was revered in the city of Oxyrhynchus, where it was never eaten, and considerable numbers of mormyrus have been found mummified in tombs in the Oxyrhynchus area. The Greek historian Plutarch tells how the fish sparked a violent confrontation between the inhabitants of Oxyrhynchus and the nearby city of Cynopolis after the people of Cynopolis, where the dog was held sacred, were one day seen eating mormyrus. The citizens of Oxyrhynchus rounded up all the dogs they could find and ate them in retaliation.

River dolphins have been venerated in several parts of the world. The South Asian river dolphin of the sacred Ganges was given religious significance with its mention in the Rig Veda and became one of the first protected species in history. It was accorded a special status under the reign of Emperor Ashoka, one of India's most famous rulers, in the 3rd century BC. In South-East Asia, both the Khmer and Lao people regard the Irrawaddy dolphin as a sacred animal, and they are rarely hunted. Likewise, in South American Indian folklore, the Amazon river dolphin is considered sacred, leading to the belief that hunting and killing them will bring bad luck. The Yangtze river dolphin was revered as the goddess of the Yangtze in China until it was declared extinct in 2007.

Various species of salmon have been honoured in myths and rituals in several societies thanks to their value as an important seasonal food source and their ability to survive in both the salty ocean and freshwater rivers. The Atlantic salmon occupied a special place in Celtic mythology. It was said to be as old as time and to know all things, both in the past and in the future. The Salmon of Wisdom, from Irish legend, features in an important episode in the early life of Fionn mac Cumhaill (anglicized to Finn McCool), a legendary hunter-warrior. Fionn was studying under a poet who had sought the fish for seven years. When the poet finally caught the fish, he asked Fionn to cook it for him, but Fionn burned his thumb on the fish and instinctively put it in his mouth to suck the burn, hence receiving all the knowledge in the world.

During the Middle Ages in Britain, salmon were known as the all-knowing water creatures in Arthurian legends. Gwrhyr, one of King Arthur's finest men and an expert linguist, talked to a series of wise animals in his search for a master huntsman. Each animal was wiser than the previous one, and the oldest and wisest of them all was the salmon of Llyn Llyw, a mythological pool on the River Severn. The magic salmon was said to have gained the power of wisdom by consuming hazelnuts that had dropped into its pool. According to this tradition, the number of spots on a salmon's back is supposed to represent the number of nuts consumed.

Salmon has long been a principal source of food for indigenous groups in northern latitudes, such as on the Pacific coast of North America before the arrival of European colonists. As such a mainstay of the diet, the fish became a focus for numerous rituals, taboos, and mythological stories. Catching the salmon as they ran up the rivers in enormous numbers to spawn was a time of great abundance, often marking a stark contrast to the times when only dried meat and fish were eaten, so the start of the season was a time for reverence and celebration. In the early decades of the 20th century, anthropologists working in the region recorded details of the 'first salmon ceremony' held among many Native American groups, to mark the initial salmon run of the season, before the practice ceased.

Among the Tsimshian communities along the Skeena River in today's Canadian province of British Columbia, any fisherman landing the first salmon was obliged to call four shamans who arrived to take charge. The fish was placed on a mat made of cedar bark and carried to the chief's house in a procession led by one of the shaman – who had put on the fisherman's clothing – holding a rattle in his right hand and an eagle's tail in the left. At the house, in the presence of senior members of the community, the shamans would march around the salmon four times before the man wearing the fisherman's clothes called for the fish's head to be severed, followed by its tail, and the removal of its stomach. The ceremony, marked by the chanting of honorary names, was conducted with a mussel-shell knife. It was thought that using a stone or metal knife would cause a thunderstorm.

Similar first salmon ceremonies were conducted up and down the Pacific seaboard with only minor differences. Some involved speeches and feasting, others ceremonial dances. All stressed respect for the salmon in the hope that it would come in great numbers. Salmon was eaten fresh during the fishing season and dried or smoked for the winter food supply. Many groups in North America, Siberia, and northeastern China also used salmon skin to make their clothing.

River spirits

The association between rivers and various mythical creatures is common to numerous cultures all over the world. In Germanic and Nordic folklore, such water sprites are known as 'nixie' (singular nix) and usually have evil intentions. They frequently entice their human victims to join them, luring them into the water, from which there is no escape. The nix may take different forms, either male or female. One of the best known from Germany was Lorelei, a beautiful nymph who sat on a rock in the Rhine which now bears her name, and lured fishermen into danger with the sound of her voice or by combing her hair. Scandinavian nixie were more likely to be male, drawing their female victims into a river or lake with enchanting songs played on the violin. Pregnant women and unbaptized children were especially vulnerable.

Another form of river spirit in Scandinavian folklore is the bäckahästen, or 'brook horse', a majestic white beast that would commonly appear on river banks, especially during foggy weather, presenting a tempting ride for a weary traveller. Anyone who climbed onto its back would be unable to dismount, enabling the horse to jump into the river and drown its rider. The kelpie of Scottish folklore is a direct parallel of the bäckahästen. Its most common guise was that of a fine-looking tame horse, but the kelpie could also appear as a hairy man with a terrible vice-like grip. He would hide on the river bank until an unfortunate traveller was passing and then leap out to crush the life from him.

A spirit associated with rivers all over Japan is the kappa, a mischievous creature often described as something between a child and a monkey. One of the kappa's favourite tricks is also to lure people, horses, or cattle into a river to drown. There are numerous regional variations to the kappa and its behaviour, but one of its most common traits is an affinity for cucumbers (frequently thought of as a symbol of fertility). In some parts of Japan, it is believed that anyone eating a cucumber before swimming will certainly be attacked by a kappa, although in other areas it is a way to ensure protection against kappa attack. Either way, many festivals associated with kappa include offerings of cucumbers, and the link between the kappa and the cucumber continues in modern Japan through the name of a type of sushi made with cucumbers: kappa maki. Interestingly, the character of the centuries-old kappa has been subject to a make-over in the last hundred years or so, and has been transformed from a malicious and unpleasant water deity into a harmless and endearing mascot. As a nationally recognized symbol, the kappa has been used for various campaigns that draw on a nostalgia for Japan's rural past. It is ironic to note that one of these was a clean water campaign aiming to regenerate the environment around urban rivers, calling for rivers to be cleaned up so that kappa will come back.

Traditionally, the kappa, like the nix, bäckahästen, and kelpie, are malevolent river spirits, luring the unwary to a watery death. In southern Africa, by contrast, the spirits associated with river systems and other water bodies in the traditional cosmologies of Khoisan- and Bantu-speaking indigenous peoples behave rather differently. To many of these groups, water spirits are regarded as ancestors and they prefer to live in certain spots. In rivers, these are deep pools, frequently below waterfalls where the water is fast-moving and 'living', often generating lots of foam. These spirits take on various zoomorphic manifestations, primarily the snake and the mermaid. They interact with humans in a variety of ways, and one of the most important of these is their fundamental importance to traditional healing and its practitioners.

Water spirits traditionally call certain chosen individuals to become diviners or healers, which usually involves the physical submersion of the candidate under the water of a certain river pool for a few hours, days, or even years. When the man or woman emerges from the depths, he or she is wearing a snake and has acquired psychic abilities and healing skills, including knowledge of medicinal plants. This experience of being taken under the water can occur in a dream, but this is simply notification that the ancestors are calling the individual to become a healer. The calling frequently comes after a period of illness, although when children are called, they often just happen to be playing near the river at the time. Resistance to this calling is not advised and usually leads to misfortune. Relatives are not allowed to display any grief at the disappearance of one who has gone under the water or the individual may never be returned.

Given the importance of the water spirits, many rivers, pools, and water sources are viewed with a mixture of awe, fear, and respect by indigenous communities in southern Africa. Their sanctity has generated numerous taboos surrounding access and use. Frequently, only healers, kings, and chiefs are allowed to approach such areas. The general populace is forbidden to go near sacred pools for fear of being taken under the water, never to return. Such taboos represent just one small fragment of the great web of influence rivers exert on humankind, a power that can be traced back to the beginnings of humanity. The ways in which the flow of rivers has helped to shape history is investigated in more detail in the following chapter.

Chapter 3
Liquid histories

I have seen the Mississippi. That is muddy water. I have seen the St Lawrence. That is crystal water. But the Thames is liquid history.

John Burns (1858–1943)
(British politician)

Rivers reflect history. They also help to create it. Societies interact with rivers for many reasons, and these motivations can be classified simply into those rooted in the useful aspects of rivers and those that reflect rivers as hazards. People derive many benefits from rivers. We have caught and eaten fish from them for tens of thousands of years. Rivers provide water for domestic, industrial, and agricultural uses. They also provide all sorts of minerals, ranging from gold and diamonds to the sand and gravel that is so essential to modern construction. The energy in a flowing river can be harnessed to facilitate trade and travel, to generate electricity, and to remove many types of waste produced by human activities. People enjoy rivers for recreational and aesthetic reasons, and as havens for wildlife. Conversely, rivers can breed fear and trepidation. This may be a function of quantity: either too much water – a flood; or too little – a water shortage. The quality of river water may also be a cause for concern, as a bearer of disease or dangerous concentrations of minerals such as arsenic.

All of these facets of rivers as perceived by society have inevitably had some bearing on the course of human history. Much of Europe's story can be told through the story of the Danube. The rivers of Bangladesh compose both the landscape of the nation and the life of its people. London is nothing without the Thames. Rivers are an essential part of the very fabric of many societies and their histories.

The first civilizations

Ancient civilizations emerged on the floodplains of large rivers in several parts of the world between 3,500 and 5,500 years ago. The appearance of the Sumerian, Egyptian, and Harappan civilizations along the alluvial valleys of the Tigris-Euphrates, the Nile, and the Indus respectively was due in large part to the key benefits offered by their rivers: an abundant supply of fresh water, fertile alluvial soils, and a ready transport corridor for trade and travellers. In each case, the society's reliance on its river was emphasized by the arid location, making inhabitants particularly dependent upon a reliable flow of water for agriculture and their continued existence in a desert environment. All three river systems are exotic: rising in areas with more humid climates which maintain their perennial flow through the desert.

One theory linking many of the factors involved in the emergence of these first civilizations suggests that the central organization required to manage irrigation in desert areas also allowed complex societies to evolve as large numbers of people congregated to live in the same place. This tendency led eventually to the creation of the first cities and what are popularly thought of as the first civilizations. Each of these three early river-based civilizations developed its own ways of diverting and channelling water, growing and storing food. Systems for writing, making laws, and many other hallmarks of civilization also emerged separately in these three regions. This theory of 'hydraulic civilizations' suggests that the deliberate manipulation and regulation of their rivers by these early societies was an inherent and necessary precondition of civilization.

Another idea takes the links between these early complex societies and their rivers a step further, to suggest that the nature, character, and longevity of the civilization was in part a reflection of the nature of its river. The Tigris-Euphrates, Nile, and Indus are all large, exotic river systems, but in other respects they are quite different. The Nile has a relatively gentle gradient in Egypt and a channel that has experienced only small changes over the last few thousand years, by meander cut-off and a minor shift eastwards. The river usually flooded in a regular and predictable way. The stability and long continuity of the Egyptian civilization may be a reflection of its river's relative stability. The steeper channel of the Indus, by contrast, has experienced major avulsions over great distances on the lower Indus Plain and some very large floods caused by the failure of glacier ice dams in the Himalayan mountains. Likely explanations for the abandonment of many Harappan cities, including Mohenjo Daro, take account of damage caused by major floods and/or the disruption caused by channel avulsion leading to a loss of water supply.

Channel avulsion was also a problem for the Sumerian civilization on the alluvial plain called Mesopotamia – 'the land between two rivers' – known for the rise and fall of its numerous city states. Most of these cities were situated along the Euphrates River, probably because it was more easily controlled for irrigation purposes than the Tigris, which flowed faster and carried much more water. However, the Euphrates was an anastomosing river with multiple channels that diverge and rejoin. Over time, individual branch channels ceased to flow as others formed, and settlements located on these channels inevitably declined and were abandoned as their water supply ran dry, while others expanded as their channels carried greater amounts of water.

Pathways for exploration

The straightforward pathways offered by rivers have always been used by people arriving to explore new lands. Archaeological evidence suggests that early humans penetrated an island later to become known as Britain along its major rivers during the

Palaeolithic period or early Stone Age, later spreading out and settling areas further from the river banks. Similarly, some 6,000 years ago, Neolithic tribes used river courses to enter Central Europe from the southeast. In both cases, river valleys offered plentiful supplies of essential resources for these early settlers: water, fish, and floodplains rich in game for hunting.

Many hundreds of years later, the great river systems of North America enabled European pioneers to explore vast new territories, opening them up to trade and eventual colonization. In the 16th century, a succession of French traders, explorers, and missionaries were the first Europeans to arrive in the Great Lakes region, following the exploration of the St Lawrence River by their fellow countryman Jacques Cartier in the 1530s. Dispatched by the king of France, their main purpose was to chart the river systems as highways that allowed access to a new continent. These were often the only thoroughfares through the otherwise impenetrable forests of North America, traversable by canoe as liquid highways or dogsled when many tributaries froze over during the winter.

By 1804, when Meriwether Lewis and William Clark were sent by US president Thomas Jefferson to explore, survey, and document an immense swathe of North America that he'd just bought from the French – the Louisiana Purchase – rivers were still the easiest routes to follow. Lewis and Clark's expedition took them up the Missouri River, across the Rocky Mountains, and down the Columbia River to the Pacific Ocean. Their expedition and the information they brought back, particularly about the Pacific northwest, played a pivotal part in the westward thrust of US expansion.

Movement along rivers played a similar role in the expansion of Russian power and influence over Siberia and the penetration of Africa by Western European powers. The importance of rivers as pathways for colonial exploration is not simply a subject of historical interest. During the colonization of the Americas in the mid-18th century and the imperial expansion into Africa and Asia in the late 19th century, rivers were commonly used as boundaries because they were the first, and frequently the only, features mapped by European explorers. The diplomats in Europe who negotiated the allocation of colonial territories claimed by rival powers knew little of the places they were carving up. Often, their limited knowledge was based solely on maps that showed few details, rivers being the only distinct physical features marked. Today, many international river boundaries remain as legacies of those historical decisions based on poor geographical knowledge because states have been reluctant to alter their territorial boundaries from original delimitation agreements.

Australia's Murray River

The Murray River in the southeast of Australia is of immense cultural, economic, and environmental importance to the continent. Its significance becomes greater still if its two largest tributaries, the Murrumbidgee and Darling Rivers, are included. Altogether, the Murray-Darling Basin drains about 14% of Australia's total land area. Numerous Aboriginal peoples relied on the abundance of the river for thousands of years before the arrival of Europeans, hunting and trading along the Murray in canoes cut and shaped from the bark of gum trees growing on the river's edge. Rock art, archaeological and burial sites remain as evidence of these early inhabitants. Their diet from the river was varied, including fish, crayfish, mussels, frogs, turtles, and waterfowl and their eggs.

It was not until the 1820s that European explorers first saw the river. Captain Charles Sturt navigated down the Murrumbidgee, followed the Murray to discover the Darling confluence, and continued downriver to the mouth of the great river. Publication in London of Sturt's account of his river exploration led indirectly to the establishment of the colony of South Australia. Early

European settlers began to penetrate the continent's interior by following the Murray and small settlements and sheep farms started to spring up along its banks. One of the most memorable symbols of the European history of the Murray River is the paddle-steamer, numbers of which ferried wool, wheat, and other goods up and down the river system, helping to open up the Murray-Darling Basin. Irrigated agriculture began in 1887, accelerating settlement and exploitation of the river's water supplies.

Today, the Murray-Darling Basin is Australia's most important agricultural area, producing over one-third of the nation's food supply. It contains 65% of the country's irrigated farmland and supports more than one-quarter of the national cattle herd and nearly half of its sheep. It also provides water to major cities including Canberra and Adelaide. In its natural state, however, the River Murray was a highly variable and unpredictable source of water. During severe droughts, it ceased to be a river at all and was transformed into a chain of salty waterholes, but flow has been regulated for many years to maintain a reliable supply. River regulation on the Murray has been achieved with an array of water engineering structures and techniques. They include five main water storage points, including two large dams – Dartmouth and Hume – and the major managed lakes of Mulwala, Victoria, and Menindee. Since the Hume Dam was completed in 1936, a continuous flow has been maintained throughout the length of the river. The Murray and the Murrumbidgee Rivers also receive additional water supplies diverted through a series of tunnels and pipes from the Snowy River. A system of thirteen weirs and locks further aids flow regulation, and five barrages have been constructed near the river mouth to prevent the intrusion of sea water. Salt occurs naturally in the Murray-Darling Basin in large quantities, and is a water-quality issue for domestic and agricultural use. Hence, a series of salt-interception schemes has been established to keep salt out of the river. These schemes involve large-scale groundwater pumping and drainage projects that intercept flows of saline water and dispose of them by evaporation.

Natural barriers

There are many examples of rivers acting as natural barriers to interaction between groups, and in some cases the separation has continued over periods long enough to be apparent in genetic studies. Among primates in Central Africa, the Congo River forms a clear divide between bonobos, or pygmy chimpanzees, which are found only on the south side of the Congo, and common chimpanzees, which occur only to the north. Chimpanzees are not known to swim, so the river has effectively isolated the two groups, for about 1.3 million years according to genetic analysis, which also confirms that they have a common ancestry.

Similar river barriers to the flow of genes, cultures, and languages have been identified in some human societies. A classic case has been documented by anthropologists in the Highlands of New Guinea where the Lamari River marks a very sharp cultural and linguistic divide between the Fore and the Anga. These two groups speak completely unrelated languages and have markedly different cultures. They are also mortal enemies. Although there may be several reasons for their differences, the formidable natural barrier presented by the Lamari River and its precipitous valley is certainly an important one, particularly given that the Fore believe people to be incapable of swimming.

Rivers have always demarcated such boundaries, both real and imagined. The River Danube in Europe marked the northern frontier of the Roman Empire in the 1st century AD because it was easily defended, hence also drawing a perceived line between the 'civilized' Empire and the barbarian tribes on the bank opposite. In Europe today, the Danube marks the international border between Slovakia and Hungary, and stretches of Romania's borders with Serbia, Bulgaria, and Ukraine. In southern Germany, the Danube is affectionately known to most Bavarians as the 'Weisswurst equator' (literally, the 'white sausage equator', named after a favourite food from the south), the symbolic borderline between themselves and the different cultures to the north.

The importance of rivers as natural barriers is reflected in the fact that no less than three-quarters of the world's international boundaries follow rivers for at least part of their course. However, rivers are also notoriously erratic boundaries thanks to their inherent tendency to move, a propensity that can result in a multitude of legal, technical, and managerial challenges for rival states on opposing banks.

These are challenges both in terms of identifying a definitive line in a dynamic natural feature and of managing a divided transboundary water resource. The two elements are of course related: deciding where exactly the boundary runs affects legal rights to the water itself and how it is used (e.g. for navigation) or abused (e.g. by pollution). Often, a river boundary follows the 'thalweg', the deepest channel in the river, but other principles are also used. Some boundaries follow the median line between the banks, or lines drawn between turning points. Others follow one of the river banks. On occasion, two countries may favour two different legal principles for determining the position of the border, often for opportunistic reasons, and the resulting dispute frequently has to be resolved by international adjudication. Even after the border has been agreed, erosion and sedimentation can alter the banks, the median, or the thalweg, to the benefit of one country and the detriment of the other.

The International Boundary and Water Commission between Mexico and the USA is a long-standing example of the interrelationship between river boundary identification and river management. The commission was created in 1884 to demarcate the border on the ground and to identify its position in the Tijuana, Colorado, and Rio Grande Rivers, but in 1944 it was also given responsibility for allocating the water resources of the Rio Grande. Today, the International Boundary and Water Commission spends most of its time on water management and allocation, rather than on boundary definition. Not all river boundary disputes are settled by peaceful negotiation, however. In 1969, a fierce conflict broke out between the Soviet Union and China lasting several months over their international boundary along the Ussuri River and specifically over the ownership of Chenpao island.

River rights and conflicts

The importance of fresh water as a resource, allied to its uneven geographical distribution in rivers, lakes, and underground aquifers, has inevitably led to political wrangling over the rights of different groups to use water. On occasion, disagreement over rights to shared water resources can lead to militarized confrontation, and the notion that so-called 'water wars' may become a leading source of conflict in the 21st century has become quite widespread in some academic and journalistic circles, as well as in political rhetoric.

Many rivers flow across (as well as along) the borders between nation states, and approximately 60% of the world's fresh water is drawn from rivers shared by more than one country. Some of the world's larger river basins are shared by a great number of countries. The Danube is the greatest of them all in this regard, its basin being shared by no fewer than nineteen countries in Europe. Five other basins – the Congo, Niger, Nile, Rhine, and Zambezi – are shared by between nine and eleven countries. These facts suggest the scale of possible river rights issues, although multiple stake-holders are by no means necessary for political discontent. The River Ganges, for example, flows through just two countries yet was the subject of a twenty-year confrontation between India and Bangladesh following completion of India's Farakka Barrage in 1975. Bangladesh complained that it was being deprived of water it could use for irrigation and was subject to increasing salinity problems thanks to diversions of water by the barrage, sited some 18 kilometres upstream from its border with India.

The dispute over the Ganges, which eventually resulted in the signing of a water-sharing accord in 1996, is not atypical. A downstream state's objection to pollution, the construction of a dam, or excessive irrigation by an upstream state, actions which will decrease or degrade the quality of water available to the downstream state, are all classic grounds for disagreement over a cross-border river. Many of these disputes are peacefully settled by international treaty, but many are not. Further, not all international treaties are designed to address all players in an international river dispute. An example here can be quoted from the Nile. Egypt and Sudan have an international agreement that governs the volume of Nile water allowed to pass through the Aswan High Dam, but none of the other eight Nile Basin countries have agreements over use of the Nile's waters. Given that Egypt and Sudan are the last two states through which the river flows before entering the Mediterranean, similar agreement over water rights with countries further upstream would seem desirable. Hammering out the details of such a treaty has, to date, proved insurmountable, and differences in opinion over rights to Nile water continue to underlie many of the political issues in this part of the world.

Agreements to resolve disputes over water resources have a very long history. The beginnings of international water law can be traced back at least to 2500 BC, when two Sumerian city states – Lagash and Umma – reached an agreement to end a dispute over the water resources of a tributary of the River Tigris in the Middle East. Wrangles over water are still a significant potential source of conflict in the Tigris-Euphrates Basin due to a lack of agreements in the contemporary era. While there is currently a water surplus in this region, the scale of planned developments has raised concerns. The Southeastern Anatolian Project in Turkey, a regional development scheme on the headwaters of the two rivers, envisages the eventual construction of 22 dams. In 1990, when the reservoir behind the Ataturk Dam began to fill, stemming the flow of the Euphrates, immediate alarm was expressed by Syria and Iraq, despite the fact that governments in both countries had been alerted and discharge before the cut-off had been enhanced in compensation. Full development of the Southeastern Anatolian Project, expected by about 2030, could reduce the flow of the Euphrates by as much as 60%, which could severely jeopardize Syrian and Iraqi agriculture downstream. The three Tigris–Euphrates riparians have tried to reach agreements over the water use from these two rivers, and the need for such an agreement is becoming more pressing.

Force has been used in conflicts over scarce water resources elsewhere in the Middle East. Attempts to divert water from the Jordan and Yarmuk Rivers led to multiple military incidents between Israel, Syria, and Jordan in the 1950s and 1960s. In 1967, just before the Six-Day War between Israel and its Arab neighbours, then prime minister Levi Eshkol declared that 'water is a question of survival for Israel', and that Israel would use 'all means necessary to secure that the water continues to flow'.

Since then, the spectre of 'water wars' has assumed greater prominence in popular views of how relations between states sharing a river basin, particularly those in the Middle East, will develop in the future. However, not all authorities see more inter-state conflicts as either inevitable or even the most important aspect of transboundary river management. National economic development is just one dimension of 'water security', the idea of sustainable access to adequate quantities of water, of acceptable quality, for multiple uses. Such uses also include social and cultural needs but important ecosystem functions too. All of these users should have rights to a river and hence conflicts between them can emerge at levels other than the nation state.

The Mekong Basin illustrates this point. Like many of the world's international river basins, it is simultaneously viewed as an engine of regional economic development, as a crucial basis of livelihood resources, and also as a vital site for the conservation of biodiversity. Critics of the 1995 Mekong Agreement between four of the basin's nation states see the treaty as overly focused on the Mekong's huge hydroelectric potential and capacity to store water for irrigation schemes. Development of this potential is inevitably concentrated at the national level, often with assistance from international development partners such as banks and other governments. These bodies, it is argued, view the Mekong's resources as under-utilized and ripe for development, but the fear is that this stance will marginalize the activities of local resource-users who depend on the river for sustenance and livelihoods.

Muddy history

The sediments carried in rivers, laid down over many years, represent a record of the changes that have occurred in the drainage basin through the ages. Analysis of these sediments is one way in which physical geographers can interpret the historical development of landscapes. They can study the physical and chemical characteristics of the sediment itself and/or the biological remains they contain, such as pollen or spores. In some places, the sediments may be exposed in a free-face – naturally, such as a cliff, or thanks to human action – and can be examined and sampled fairly easily, but in most cases the sequence of sediments is sampled from the top down, back through time, using a device that drills a core.

The simple rate at which material is deposited by a river can be a good reflection of how conditions have changed in the drainage basin. For example, a study of sediment laid down over a period of 300 years by the Bush River, which flows into Chesapeake Bay on the eastern seaboard of North America, has shown that the amount of soil eroded from the catchment has altered significantly in response to changing land use in the area. Before the settlement of Europeans in the Bush River basin, which began in the mid-17th century, native populations are thought to have had no significant environmental impact on the basin, and the sedimentation rate before 1750 was about 1 millimetre per year. However, the rate was eight times greater by 1820 thanks to early deforestation and agriculture practised by the first Europeans. As the felling of trees progressed and agriculture intensified over the next 100 years, greater erosion followed, and sedimentation rates peaked in 1850 at about 35 millimetres per year. In more recent times, since 1920, urbanization – protecting soils – and the building of dams – blocking the delivery of sediment – have combined to reduce erosion and sedimentation by an order of magnitude. The sedimentation rate has been reduced nearly to the background conditions that prevailed in the pre-European settlement era.

Pollen from surrounding plants is often found in abundance in fluvial sediments, and the analysis of pollen can yield a great deal of information about past conditions in an area. The type of vegetation can be modified by all sorts of factors, including human interference as in the Bush River basin example, but also for entirely natural reasons such as a change in climate or soil conditions. Very long sediment cores taken from lakes and swamps enable us to reconstruct changes in vegetation over very long time periods, in some cases over a million years and more. Because climate is a strong determinant of vegetation, pollen analysis has also proved to be an important method for tracing changes in past climates.

An important study of a 250-metre-long core from the bed of Lake Biwa in Japan, for instance, showed changes in pollen over about the last 430,000 years, a period in which five glacial–interglacial cycles could be recognized. During glacial periods, pollen from pine, birch, and quercus (or white oak) trees was dominant, indicating a climate that was cool and temperate, tending towards subarctic. During interglacial periods, by contrast, high pollen values for species typical of a warm–temperate climate were found, including broad-leaved trees such as the deciduous Lagerstroemia (or Crape myrtle) and the evergreen Castanopsis (a type of beech).

Other evidence for environmental change can be detected in larger-scale elements of the landscape. The floodplains of many modern rivers, for example, bear traces of former channels, so-called 'palaeochannels', which are different in scale and/or form from the current river. If a palaeochannel is buried beneath more recent sediments, it was probably formed by a river that flowed towards a lower base level, indicating that local sea level or lake level has subsequently changed. River terraces that are found in many river valleys are thought to reflect fluctuations of climate, their formation having been driven by the direct and indirect influence of temperature and precipitation on fluvial activity.

Water power

The energy in flowing and falling water has been harnessed to perform work by turning waterwheels for more than 2,000 years. The moving water turns a large wheel and a shaft connected to the wheel axle transmits the power from the water through a system of gears and cogs to work machinery, such as a millstone to grind corn. An early description of a water-powered mill for grinding grain is given by a Roman engineer named Vitruvius, who compiled a treatise in ten volumes covering all aspects of Roman engineering, and the eastern Mediterranean is strongly associated with the first use of this technology, although a separate tradition of using water power also emerged at about the same time in China. Roman waterwheels were frequently connected to other forms of hydraulic engineering, such as aqueducts and dams, designed to transport river water and control its flow to the wheels. Multiple sets of Roman watermills for grinding grain into flour on a large scale are known from the Krokodilion River near

Caesarea Maritima in today's Israel, and from Chemtou and Testour on the River Medjerda (the ancient Bagradas River) in the Roman cornbelt of North Africa, part of Tunisia today. The mills just outside the town of Caesarea Maritima consisted of four vertical waterwheels fed by an aqueduct from a dam on the river.

The power of rivers became widely used in the ancient world for milling grain but also for other purposes. Water-powered mills were also developed to drive trip-hammers for crushing ore and saws for cutting rock. All sorts of water-powered machines became more and more common in medieval Europe, gradually taking over tasks from manual labour. The early medieval watermill was able to do the work of between 30 and 60 people, and by the end of the 10th century in Europe, waterwheels were commonly used in a wide range of industries, including powering forge hammers, oil and silk mills, sugar-cane crushers, ore-crushing mills, breaking up bark in tanning mills, pounding leather, and grinding stones. Nonetheless, most were still used for grinding grains for preparation into various types of food and drink. The Domesday Book, a survey prepared in England in AD 1086, lists 6,082 watermills, although this is probably a conservative estimate because many mills were not recorded in the far north of the country. By 1300, this number had risen to exceed 10,000.

All across Europe, the watermills generally belonged to lords, to city corporations, or to churches or monasteries. Cistercian monasteries were instrumental in the initial development in England of the 'fulling' mill in the late 12th century. Fulling, or felting, was one in a sequence of processes during the production of woollen cloth produced on the monastic estates. It involved scouring and consolidation of the fibres of the fabric, both necessary for proper finishing. The introduction of water-powered technology revolutionized fulling, a process that had hitherto relied on human power to beat the cloth. On the Isle of Wight in southern England, for instance, the first fulling mill was established at the Cistercian monastery of Quarr Abbey on a stream close to large areas of pasture on the abbey's estates. Wool from the flocks of sheep was processed at the abbey and sold in nearby towns.

Medieval watermills typically powered their wheels by using a dam or weir to concentrate the falling water and pond a reserve supply. These modifications to rivers became increasingly common all over Europe, and by the end of the Middle Ages, in the mid-15th century, watermills were in use on a huge number of rivers and streams. The importance of water power continued into the Industrial Revolution, when a series of inventions transformed the manufacture of cotton in England and gave rise to a new mode of production: the factory system. The early textile factories were built to produce cloth using machines driven by waterwheels, so they were often called mills.

The supremacy of running water was soon superseded by steam power generated by burning charcoal, coal, and later oil and gas, although rivers have continued to play a role in industrial power generation. All thermal electric generating stations, whether the source of heat they use is fossil fuels, nuclear, or geothermal, convert water – or some other fluid – into steam to drive electricity-generating turbines. The steam has to be condensed in a cooling system in order to be recycled through the turbines, and large quantities of water are also required for this purpose. Much of this water is drawn from rivers, along with lakes and aquifers and the oceans.

The energy potential of water moving in rivers has re-emerged in the modern era with the advent of hydroelectricity generation. Hydropower is the only renewable resource used on a large scale for electricity generation, and about one-third of all countries rely on hydropower for more than half their electricity. Globally, hydropower provides about 20% of the world's total electricity supply. Most large hydroelectric stations rely on a dam to supply a

8. Cotton mills on the River Irwell in Manchester, northern England, c. 1850, where flowing water in rivers and canals was a crucial component of the Industrial Revolution

reliable flow of water to turn their turbines, but small 'run of river' hydroelectric stations do not need such obstacles to the natural flow of the river. Countries with abundant rainfall and mountainous terrain have developed hydropower to become their foremost supply of electricity. Norway is an interesting example.

Its rivers provide more than enough hydroelectricity for its own needs so the country has become an exporter of hydropower.

Trade and transport

The flow of water in a river has always provided another obvious utility for people: as a conduit for travel, trade, and transport. Many of the world's major cities have developed on navigable rivers thanks to the access they offer, to and from their terrestrial interiors and, in many cases, to territories overseas. The River Thames and London provide a good example. In medieval England, the transportation of goods along the river played an important role in the development of London as a city, and indeed of many other settlements in the Thames valley. Water transport was attractive at this time because of its relatively low cost: moving commodities such as grain and wool by land could be more than ten times the price of transport by water. Cheap transport by river stimulated economic development by increasing the size of markets, encouraging regional specialization, and promoting urbanization. Historical research of transport along the Thames and its tributaries around the year 1300 shows that these waterways greatly extended the market for grain and fuel supplied to the capital. The specialization in farming that developed around London at this time is also likely to have been a result of the increase in transport by river, since some areas were better suited to the production of particular crops than others. Two main impacts on urban development can be identified. For London, development of the cheap fluvial transport network removed a constraint on the city's expansion because it reduced the cost of food and fuel used in the capital. Urban development was also stimulated outside London, in the capital's hinterland, as towns such as Henley-on-Thames grew to become a specialized centre supplying agricultural produce to the city.

Navigable rivers also became major arteries of trade and stimulated the growth of larger settlements elsewhere in medieval

England. Gloucester and Bristol were served by the River Severn, York had quays on the River Ouse, and Norwich on the River Wensum. The importance of water transport to urban development was even embodied in the early 12th-century Laws of Edward the Confessor, an Anglo-Saxon king of England. The laws note that navigation should be maintained on the major rivers 'along which ships transport provisions from different places to cities or burghs'.

Many economic historians suggest that England's rivers provided the cheapest form of inland transport for hefty goods right up until the 18th century. Nonetheless, bargemen and merchants wanting to use rivers for trade in the Middle Ages had to struggle constantly against those who wanted to build mills and fish-weirs. The mid-18th century is thought of as the birth of the 'canal age' in England when industrialists built their own waterways, an era that followed a 150-year period in which water transport became progressively easier as many of the country's rivers were 'improved'.

The importance of river transport has played a key role in the economic development of many countries. In Sweden, for example, in the 17th and 18th centuries, logs felled in the country's northern forests were floated down rivers to the mining district of central Sweden where they were used as fuel in smelting operations. During the second half of the 19th century, this form of river transport played an important role in the industrialization of Sweden thanks to the rapid development of an export-oriented forest industry, based on sawmills and later pulp mills. Sweden was able to supply a growing demand for sawn wood and square timber in the industrially developing economies of Western Europe from her forests in the remote northern parts of the country. Timber felling was possible over large areas of northern Sweden thanks to the country's dense network of tributaries and main rivers which generally flow from north to south, enabling cheap long-distance transport of timber to the sawmills and pulp mills on the coast. Sweden's distinctly seasonal climate was also favourable to the transport of timber, since the spring thaw swells rivers with snowmelt that facilitates the floating of logs. At the beginning of the 20th century, sawn timber, pulp, and paper accounted for about half of Sweden's exports by value. The importance of rivers in moving timber within the country only waned in the 1980s when timber-floating operations were abandoned in favour of an expanding road network.

River transport of both goods and people continues to be economically important in Bangladesh which, with some 700 rivers and major tributaries criss-crossing the country, has one of the world's largest inland waterway networks. The total length of rivers navigable by modern mechanized vessels shrinks during the dry season, but it still connects almost all the country's major cities, towns, and commercial centres. Indeed, inland ports in Bangladesh handle about 40% of the nation's foreign trade.

Inland water transport is cheaper than road or rail, and is often the only mode that serves the rural poor, proving especially useful during periods of widespread flooding in the monsoon season when many roads become impassable. Country boats, the traditional mode of river transport in Bangladesh for centuries, are also the main means of transport at any time in southern areas of the country where the road network is little developed.

In some parts of the world, trade along rivers has involved contact between radically different cultures, with a range of impacts, some beneficial, others less so. In North America, the fur trade stimulated Euro-American exploration of the Missouri River Valley in the 1700s. Native American Indians met European traders at certain points along the river, some of these trade centres pre-dating contact with Europeans by hundreds of years. American Indians provided beaver pelts and buffalo hides in exchange for manufactured and processed goods such as metal cooking pots, knives, guns, fabrics, beads, coffee, and sugar. By the

9. Several hundred thousand country boats ply the rivers of
Bangladesh, moving passengers and cargo. These boats play a vital role
in the lives of rural people and in the rural economy

1800s, steamboats plied up and down the Missouri River from the
town of St Louis where commercial trading companies had
established bases. American Indians were exposed to many
aspects of Euro-American culture but also, inadvertently, to
deadly diseases to which they had no immunity. One epidemic of
smallpox in 1837, probably transferred to Plains Indian tribes by a
steamboat passenger, killed 10,000 to 20,000 Indians, including
over 90% of the Mandan Nation, proprietors of one of the
Missouri River's main trading posts.

The Danube: artery of Europe

Although not its longest, many would offer the Danube as
Europe's principal river, as it was in the mid-17th century when
Pope Innocent X approved construction of the Four Rivers
Fountain in Piazza Navona in Rome. Crowned by an Egyptian
obelisk, Gian Lorenzo Bernini's most dramatic and spectacular

work consists of four marble figures symbolizing the major world rivers known at the time (no doubt in part a reference to the four rivers of Eden). The Nile represented Africa, the Ganges Asia, the Rio de la Plata symbolized the Americas, and the Danube represented Europe. Linking more countries than any other river in the world, the Danube both defines and integrates the continent.

Human occupation of the Danube Valley has been traced back at least 25,000 years, when men gathered to hunt mammoth at Dolni Vestonice, in the Czech Republic of today. A natural corridor for migration linking east and west, the river was used by farmers from the Anatolian Peninsula seeking new lands to cultivate some 7,000 years ago. Five millennia later, the Persian King Darius led his vast army along the same route before crossing the Danube in his campaign against the Scythians. The river was established as a corridor for trade by the ancient Greeks and during Roman times, when the Danube was used both as a defensive barrier and as a supply line to feed and equip the legionnaires stationed along it.

Europe's Christian military forces used the Danube as a pathway for heading towards Byzantium and the Holy Land during the times of the Crusades a thousand years ago, and in the 16th century the Danube provided the route for a reverse crusade when Suleiman the Magnificent brought Islam westwards from the Black Sea. In the 1520s, the Ottoman Turks took Belgrade, defeated Hungary, and advanced to the walls of Vienna. They held Budapest for 150 years before being driven back down the Danube.

Trade along the Danube gave rise to two major empires, the Austrian and Hungarian, which merged under the Habsburgs, known to German-speakers as the *Donaumonarchie*, or 'Danube Monarchy'. Maria Theresa, archduchess of Austria and queen of Hungary and Bohemia, founded an imperial government department to oversee navigation on the river. Today, the 'prince of all European rivers', as Napoleon Bonaparte liked to call the Danube, flows through four of the continent's capital cities (Vienna, Bratislava, Budapest, and Belgrade) and through, or along the borders of, ten countries. Its role as an important artery for European trade has continued, and navigation along the entire river has been promoted since the first Danubian Commission was set up in the 19th century. The International Commission for the Protection of the Danube River, established in 1998, works to ensure the sustainable and equitable use of freshwater resources in the entire Danube Basin, including the improvement of water quality and the development of mechanisms for flood and accident control.

Given the important role played by this river throughout the history of Europe, its reflection in various aspects of European culture is not unexpected. Before Bernini's Four Rivers Fountain in Rome, the Danube had spawned a school of landscape painting in the 16th century. Some 200 years later, it became the subject of a famous musical waltz by Johann Strauss the Younger. These examples serve to illustrate some of the many ways in which rivers have presented stimulation and inspiration to writers and artists, a subject examined in more detail in the next chapter.

Liquid histories

Chapter 4
Roads that move

Rivers are roads that move and carry us whither we wish to go.

Blaise Pascal (1623–62)

(French mathematician and philosopher)

Rivers have interested humankind for millennia. They feature prominently in many facets of culture, providing liquid inspiration to diverse sectors of artistic society, from poets to musicians. The currents of a river have been harnessed not only to embody the bucolic mysteries of nature but to carry ideas and motifs, and to propel writers into the past. As an ever-flowing symbol of God's work, the river combines both the spiritual and the physical, offering an insight into humanity's place in the order of things. The long history of the river's importance to literature and the arts stretches from the poetry of Virgil to the celluloid of Francis Ford Coppola.

Rivers and language

The long, rich cultural relationship with rivers has many interesting linguistic connotations. The names of numerous rivers are in themselves descriptive. The awe-inspiring scale of flow seen in some large rivers has simply resulted in them being called 'big' or 'mighty', such as the Ottawa River in Canada, which derives its name from the Algonquin word. Others are a bit more graphic. In England, the River Thames' name is believed to come from an Indo-European word meaning 'dark river'; the River Wellow was winding, the Swift fast-flowing, and the Cray was pure or clear. Names of Celtic origin abound in Britain: the River Dart is a Celtic river name meaning 'river where oak-trees grow', and the River Iwerne is thought to mean 'lined with yew trees'. Conversely, however, lots of rivers have names that simply mean 'river'. The Avon in the west of England gets its name from a Celtic word meaning river, so that River Avon literally means 'River River'. Similarly, the River Ganges in South Asia takes its name from the Sanskrit word *ganga*, meaning current or river.

Rivers have also had their names appropriated for use as place names. Cities named after their rivers include the capitals of Russia (Moscow: Moskva River), Lithuania (Vilnius: Vilnia River), Central African Republic (Bangui: Ubangi River), and Malawi (Lilongwe: Lilongwe River). Belmopan, the capital city of Belize, was named after two rivers: the country's longest, the Belize River, and one of its tributaries, the Mopan. On a still larger scale, a number of countries are named after their major rivers. They include Paraguay in South America, Jordan in the Middle East, Gambia and Senegal in West Africa. Further east in West Africa, the Niger River flows through both Niger and Nigeria, and Central Africa's Congo has given rise to both the Congo Republic and the Democratic Republic of the Congo. India is named after the Indus River, although it no longer flows through India. A country of sorts was created in northern Europe in 1806 by Napoleon Bonaparte when he established the Confederation of the Rhine, but it disintegrated after Napoleon's abdication in 1814.

Equally, numerous place names are linked to rivers in less direct ways. Oxford means a crossing place, or ford, used by oxen. Cambridge is traced back to 'Bridge on the River Granta' with the change from Grant-, a Celtic river name, to Cam- thought to be due to a Norman influence. Many names of settlements located at the mouth of a river have an equally simple etymology: Yarmouth and Falmouth lie at the mouths of the Rivers Yar and Fal. Of course, the same principle also applies in many other languages. Aberdeen, the port in northeast Scotland, has a name of Celtic origin (*'aber'*, or mouth, of the River Don, now Deen). Similarly, Aarhus, the port in eastern Denmark, simply means 'river mouth' in Old Danish (*aa*, river, and *os*, mouth). In the USA, a number of states have names derived from Native American words associated with rivers. Connecticut comes from a Mohican word meaning 'long river place'; Mississippi is thought to mean 'great river' in Chippewa; Missouri is an Algonquin term meaning 'river of the big canoes'; and Nebraska is from an Omaha or Otos Indian word meaning 'broad water' or 'flat river'. Not all place name links to rivers are reliable, however. A good example is the Brazilian coastal city of Rio de Janeiro, named by Portuguese sailors who first discovered the spot on New Year's Day 1502. They called it 'January River', thinking – wrongly – that the large bay on which Rio now stands was the mouth of a great river.

Some terms derived from rivers have been adopted for more general use in the English language. Meander is a good example; as both verb and adjective, it has entered the vernacular to indicate a winding path. The word 'rival' – someone competing with another for the same objective – also has its origin in riverine terminology. It is derived from a Latin word, *'rivalis'*, that means 'using the same stream'. The well-known phrase 'crossing the Rubicon' has its roots in history. The River Rubicon marked the boundary between two parts of the Roman Empire, and no Roman general was allowed to bring his forces south over the river because to do so was a direct challenge to the authority of Rome. Hence, when Julius Caesar decided to cross the river and march on Rome, he passed a point of no return in crossing the Rubicon.

Landscape painting

Rivers and their valleys have provided a rich source of stimulation for landscape painters in numerous parts of the world. Twisting channels wind their way through the long history of landscape painting in China. Probably the best-known painting from the Sung Dynasties, for instance, is the scroll entitled 'Along the River during the Qingming Festival' created by Zhang Zeduan in the early 12th century. Its panoramic depiction of daily life at the Sung capital, Bianjing (today's Kaifeng), is famed for its great detail of people, buildings, bridges, and boats clustered around and along the river. The painting has been mimicked by more than twenty other artists of subsequent dynasties. The most recent of these was a computer-generated animated version produced for the World Exposition in Shanghai in 2010 and shown in the Chinese Pavilion.

Some early examples of landscape painting in Europe are traced to the beginning of the 16th century, when a number of German and Austrian artists became associated with the Danube School of Landscape Painting. Based largely in the imperial city of Regensburg, their work combined Upper Italian Renaissance influences with German Gothic traditions. More than 300 years later, many of the French Impressionists drew inspiration from the transient colours and effects of light playing on the waters of the River Seine. They include Auguste Renoir, Claude Monet, Edouard Manet, and Gustave Caillebotte. Monet chose to live near the river in the village of Giverny, not far from Paris. The Seine also features in the work of later French artists, including one of Georges Seurat's best-known pointillist paintings, *A Sunday Afternoon on the Island of La Grande Jatte – 1884* (La Grande Jatte is an island in the Seine, at that time used as a bucolic retreat from the grimy centre of Paris). The Seine also provided early inspiration for the Fauvist painters Henri Matisse and Maurice de Vlaminck before they moved to the warmer climes of the Mediterranean.

Elsewhere in Europe, John Constable, the English Romantic painter of the early 19th century, is intimately associated with the River Stour particularly. Constable was born in East Bergholt, a village on the Stour in East Anglia, and the area around the river – Dedham Vale – has become known since the artist's lifetime as Constable Country. At more or less the same time, the work of the

10. *The Skiff* (*La Yole*), painted in 1875 by Pierre-Auguste Renoir. The scene is set on the River Seine, which provided a great influence for many Impressionist painters

Chernetsov brothers on the River Volga sparked a greater appreciation of landscape in Russian art (see below).

In North America, an artist named Thomas Cole made his first trip up the Hudson River to Catskill in 1825 and the paintings that resulted from this foray created a sensation in the nascent New York art world. The resulting Hudson River School lays claim to being the first coherent school of art in the USA. The group's members initially focused on panoramas along the Hudson in New York State, in celebration of the untamed landscapes, but their scope later widened to include subjects as distant as South America and the Arctic. Another US artist whose work is closely associated with a river, in this case the Mississippi, is John Banvard. In 1840, Banvard began painting large panoramas of the Mississippi which eventually culminated in a canvas some 800 metres in length (about half a mile, although it was advertised as being three miles long). Banvard put his work on display to the paying public and later took the Mississippi panorama to Europe, where he gave a private view to Queen Victoria in Windsor Castle, near London, in 1849.

The Volga: soul of Russia

Europe's longest river, the Volga, occupies a special place in the Russian psyche as a beloved symbol of national culture. Venerated in folklore, song, poetry, and painting, 'Mother River' or 'Mother Volga' represents the country's vast open spaces and embodies the lifeblood of Russia's history. The river was portrayed as a symbol of Russia in the sentimental poetry of several 19th-century writers, including Nicolai Karamzin, Ivan Dmitriev, and Nicolai Nekrasov. Prince Pyotr Viazemskii, a leading figure in the so-called Golden Age of Russian poetry during the first half of the 19th century, celebrated the Volga 'as a marker of nationality'. The lives of Volga river people were also vividly portrayed in the novels and stories of Maxim Gorky, one-time dishwasher on a Volga steamship whose early years were spent in the city of Nizhny Novgorod, at the confluence of the Volga and the River Oka.

Esteem for the Volga is a familiar focus of Russian folk songs, epitomized by the 'Song of the Volga Boatmen', a shanty traditionally sung by the river's barge-haulers who, in the era before steam, used to haul vessels along certain stretches of the river using ropes from the shore. The song was popularized by the operatic bass singer Feodor Chaliapin, himself born in the Volga region. It is intimately linked with the famous oil painting of the same name, by Ilya Repin, a striking depiction of the peasantry's terrible working conditions in Tsarist Russia, echoed in a Nekrasov poem: 'Along the river there were barge-haulers,/ and their funereal cry was unbearably wild.' Repin's work, completed in 1873, also managed to capture the dignity and fortitude of the barge-haulers, and represented a key stage in the development of the national realist school of painting. The latter half of the 19th century was a time when the river, its towns, villages, and surroundings were increasingly depicted on canvas by such celebrated Russian artists as Isaac Levitan, Ivan Shishkin, and Boris Kustidiyev. The work of Levitan particularly is known throughout Russia for its propensity to reflect the soul of Russian nature. He spent several summers on the river, and some of his best-known paintings capture the changing light, rhythm of life, and the beauty and serenity of the Volga's scenery.

Serious appreciation of the rural landscape in Russian art has been traced back to 1838, when two brothers, Grigory and Nikanor Chernetsov, were dispatched by the Ministry of the Imperial Palace under Tsar Nicholas I to travel the length of the Volga from Rybinsk to Astrakhan on a 'voyage of discovery', commissioned to draw panoramic views of 'the beautiful places on both banks of the Volga'. The result was a cyclorama some 600 metres long that was put on display in St Petersburg, in a room decorated to resemble a ship's cabin and equipped with sound effects to simulate the river journey. Sadly, the epic work did not survive the numerous unwindings of these viewings, but the Chernetsovs' journals and travel notes remain, along with some of their working sketches and oil paintings.

On film, a classic movie from the Soviet era is the musical comedy *Volga, Volga*, said to have been a favourite of the leader Joseph Stalin. The film tells the story of a talented folk singer who overcomes petty bureaucrats to travel to Moscow for a music contest and is set largely on a Volga steamboat named *The Josef Stalin*. First released in 1938, its light-hearted escapism stood in stark contrast to the economic hardships and political purges occurring in the Soviet Union at the time.

Music

Three water nymphs from the River Rhine are central characters in the monumental four-opera cycle *Der Ring des Nibelungen* (usually known in English simply as the Ring Cycle) by Richard Wagner. The Rhine maidens (nixie borrowed from Germanic folklore – see Chapter 2), are guardians of the *Rheingold*, a treasure hidden in the river which is stolen and turned into the ring at the centre of the mid-19th-century epic. They appear in the first and last scenes, eventually rising from the waters of the Rhine to reclaim the ring from the ashes of Brünnhilde's funeral pyre.

The charm and romance of the Danube is evoked in *The Waves of the Danube*, a waltz composed in 1880 by the Romanian Ion Ivanovici, but the waltz written 14 years earlier by the Austrian conductor and composer Johann Strauss the Younger is more widely acclaimed. *An der schönen blauen Donau*, better known in English as the *Blue Danube*, has been one of the most consistently popular pieces of classical music ever since.

Johann Strauss lived and worked in Vienna, then the capital of the Austro-Hungarian Empire, a centre of high culture and classical music. In Bohemia, at the time part of the empire, the Czech composer Bedrich Smetana wrote a cycle of nationalistic symphonic poems entitled *Ma vlast* (My Country), of which his portrait of the Vltava River remains the most popular piece. The musical depiction of the river's course across Bohemia flows through forests and meadows, past ruined castles and a peasant wedding, before sweeping majestically through Prague to join the River Elbe. The evocative piece cemented Smetana's position as one of the founders of the Czech nationalist movement and 'Vltava' is considered by many to be the unofficial national anthem of the Czech Republic.

The Mississippi is another river with a very strong musical tradition, particularly along its lower reaches where the river flows through that region of the USA known as the Deep South, a culturally cohesive farming area dominated by cotton plantations during the 19th and much of the 20th century. The various styles of music that originated along this part of the Mississippi have been enjoyed all over North America and beyond. The blues were created on the Mississippi Delta, the alluvial floodplains that stretch between the Mississippi and Yazoo Rivers, while further downstream the city of New Orleans gave rise to boogie-woogie and jazz. The blues became fused with gospel music to spawn rhythm and blues, rock 'n' roll, and soul music. Louis Armstrong, B. B. King, Chuck Berry, Jerry Lee Lewis, Elvis Presley, and Aretha Franklin are among the internationally renowned musicians of the 20th century born and raised on the banks of the Mississippi.

Rivers in literature

Authors and poets have used rivers in numerous ways. A river can serve not only as a geographical feature but as a literary device, its constant movement and direction giving impetus to a narrative. The river journey is one of the most common river metaphors, linking the past to the present, doubling as the journey through life, presenting insights into the experience of growing up. As a setting in fiction, the river bank offers a sense of destiny and hints at the possibility of self-discovery.

An assessment of the various ways rivers are used as poetic devices in Roman literature highlights how this vigorous and variable element of the landscape interacts with the dynamics of poetry. The river can be a mediator between poetry and poet, the flow of the river can become part of the narrative and may form part of a narrative structure. Not least in the epic *Aeneid* (19 BC) of Virgil, where the river serves as a symbol for directional progress, the journey being simultaneously spatial, temporal, and literary. The River Tiber is where Aeneas begins his travels in Italy and also provides a course for the narrative.

Another example of a river driving a poetic narrative is found in Alfred Lord Tennyson's poem *The Lady of Shalott* (1833). Everything in the poem follows the movement of the river. While the lady sits in her tower, the river reflects the world passing her by as it flows downstream to Camelot. When Sir Lancelot trots past on his horse, the lady leaves the tower and joins the reality of the river, unchaining the boat on its bank and writing her name on its prow, effectively discovering herself by establishing her identity. Her boat floats down the river to Camelot, where she dies.

Freedom, change, and metamorphosis, all qualities inherent in the course of a large river, appear clearly as themes in Mark Twain's *Huckleberry Finn* (1885), a quintessential river story set on the Mississippi. Huck Finn, the son of an abusive, alcoholic father, flees on a raft with his friend Jim, a runaway slave, down the Mississippi river. Their journey represents escape from oppression, a broken family life, racial discrimination, and social injustice, and the book draws on the author's own boyhood experiences along the Mississippi. Samuel Clemens – Twain's real name – also worked as a riverboat pilot in his twenties, an experience that gave him his pen-name, taken from a frequent call

79

made by the man sounding the depth of the river in shallow places. Relayed to the pilot in order to keep the boat from running aground, 'mark twain' meant 'by the mark two fathoms'.

The change and renewal are more fantastical in *The Water Babies* (1863), Charles Kingsley's classic children's novel, which begins with the boy Tom, a chimney sweep, seeking a river's cleansing properties. Tom escapes his terrible life to find freedom in the river but, after his adventures as a water baby, he is finally reborn in human form once more, in a moral tale of Christian redemption. One of the most powerful works of fiction centred upon an urban river, Charles Dickens' *Our Mutual Friend* was begun the year after publication of Kingsley's novel. Published in serial form, it uses the River Thames in Victorian London to bestow rebirth and renewal upon several characters and is awash with watery imagery. The Thames is used in a similar way to change identity by William Boyd in his book *Ordinary Thunderstorms* (2009), a novel he was prompted to write by learning that the police pull a dead body from the river every week on average.

In literature, rivers are also used as agents of transformation through their representation of boundaries or thresholds, so that the practice of crossing a river precipitates some sort of change. Rivers can unify or divide, act as companion or god. Embracing the essential mysteries of nature, rivers can embody the pursuit of wisdom. They can be used to explore the physical world for our moral and intellectual, as well as physical, orientation. And of course, even within a single work, a river has many meanings.

The Congo: *Heart of Darkness*

Joseph Conrad's *Heart of Darkness* is considered by many to be the ultimate 'modernist' novel, a work of great complexity designed to reflect the complexity of experience we find in the real world. The thread running through the book, Africa's Congo River, helps to lend both direction and form to its uncertainties. The story is a simple quest, an adventurous journey upriver by one man, Marlow, in search of another, Kurtz. This is a physical journey, into a continent along a river, but also a moral and political journey, confronting the harsh realities of colonialism (Kurtz is a lost agent who works for a Belgian company involved in the ivory trade). The journey also works on another level still, becoming a psychological trip, undertaken by Marlow and the reader, in which we descend into ourselves to confront our basic drives and impulses, weaknesses and needs, a descent into the underworld that is the 'Heart of Darkness'.

The book is constructed as a tale within a tale, the narrative beginning on the estuary of the River Thames, where four men sit on the deck of a ship listening to Marlow tell his story of a trip to Africa in his youth. The setting allows the implications of what happens in the 'dark places' of a far-away continent to reverberate through the seemingly safe and comfortable world of the audience.

During Marlow's voyage upriver, an image of Kurtz gradually emerges. A man who started out as a force for good has been corrupted by the exercise of power. Kurtz has acquired a status in the local African community that is almost divine, a position consolidated by his use of force: he has plundered the countryside for its ivory, shooting people at will and displaying their skulls on his picket fence as a symbol of his authority. Marlow's journey into the heart of Africa is an exploration of the shadowy underbelly of the European Enlightenment, the language of reason, and the rhetoric of imperialism.

Conrad's *Heart of Darkness* was first published in serial form in *Blackwood's Magazine* right at the end of the 19th century, and as a book in 1902. Towards the end of the 20th century, Marlow's river trip was re-enacted in another classic of fiction, this time on celluloid, in Francis Ford Coppola's spectacular film about the

11. Joseph Conrad immortalized the Congo in his classic modernist novel *Heart of Darkness*, a book also considered to have generated many condescending Western perceptions of sub-Saharan Africa

Vietnam War, *Apocalypse Now* (1979). The movie, although of course ostensibly set on another continent, showed that Conrad's story still had numerous contemporary echoes almost a century after its creation. Colonel Kurtz, a special forces commander driven insane by power, played by Marlon Brando, still represents the corrupted voice of Enlightenment, humanism, and supposed progress. The film, like the book, develops imagery and characters that can be interpreted as a searing criticism of war, racism, and colonialism. However, both book and film have also been viewed as expressions of the hypocritical values they are trying to expose.

Coppola's film also came with a tale within a tale, simultaneously generating the documentary film *Hearts of Darkness*, a record of the making of *Apocalypse Now* that was a testimony to real-life corruption, decadence, and insanity worthy of the fictional Kurtz. In all cases – the book, the film, and the film about the film – the story is told only from the perspective of the outsiders. No effort is made to understand the alien continent through which the river flows. This can be criticized as emblematic of Europe's mythologizing of Africa in general, and of the Congo in particular, and of the USA's blinkered crusade for freedom and democracy. But this lack of an alternative frame of reference is also essential to the multi-faceted objectives of each story. Each is intended to be an essentially solitary journey involving profound spiritual change in the voyager, a mission to the very centre of things that cannot find simple answers to the questions of human existence. Kurtz's character remains as enigmatic as the darkness in which he has taken up residence. In each case, the river plays a pivotal role, in Conrad's words, as a conduit for the 'dreams of men' and the 'germs of empires'.

Chapter 5
Tamed rivers

> The servitude of rivers is the noblest and most important victory
> which man has obtained over the licentiousness of nature.
>
> Edward Gibbon (1737–94)
> (English historian)

People have interacted with rivers throughout human history and their impacts, both direct and indirect, have taken many forms. The earliest examples of water being extracted from rivers on a significant scale for the irrigation of crops date back 6,000 years. Deliberate manipulation of river channels through engineering works, including dam construction, diversion, channelization, and culverting, also has a long history. Some of the world's oldest dams, in the Middle East, were built more than 4,500 years ago and deliberate diversion and regulation of the Yellow River in China, for example, began more than 2,000 years before the present. Since these early examples, the deliberate human alteration of rivers all over the world has expanded in its extent and escalated in its ambition and scale. Nevertheless, significant geographical differences in the degree and intensity of river modifications remain. In Europe today, almost 80% of the total discharge of the continent's major rivers is affected by measures designed to regulate flow, whether for drinking water supply, hydroelectric power generation, flood control, or any other reason. The proportion in individual countries is higher still. About 90%

12. Rivers in even the most inaccessible regions are affected to some extent by human activities, as here in the Darien Gap in Panama, an area renowned globally for its remoteness

of rivers in the UK are regulated as a result of these activities, while in the Netherlands this percentage is close to 100. By contrast, some of the largest rivers on other continents, including the Amazon and the Congo, are hardly manipulated at all.

Direct and intentional modifications to rivers are complemented by the impacts of land use and land use changes which frequently result in the alteration of rivers as an unintended side effect. Deforestation, afforestation, land drainage, agriculture, and the use of fire have all had significant impacts, with perhaps the most extreme effects produced by construction activity and urbanization. These impacts are diverse and are not all direct. Many aspects of a dynamic river channel and its associated ecosystems are mutually adjusting, so a human activity in a landscape that affects the supply of water or sediment is likely to set off a complex cascade of other alterations. When contemporary climate change is included in the vast array of human activities that in some way result in changes to rivers, many authorities argue that few, if any, rivers – even in the world's least populated regions – remain unaffected by human impact. In many ways, therefore, the evolution and development of rivers is driven as much by social and economic factors as by natural ones.

Irrigated agriculture

One of the most important developments in human society was the shift from a subsistence way of life based on hunting and gathering food from the wild to one primarily based on food production derived from cultivated plants and domesticated animals. The links between early agricultural management and the emergence of urban civilizations in just a few independent centres around the world have been noted in Chapter 3 along the alluvial valleys of the Tigris-Euphrates, the Nile, and the Indus. These links developed from the high levels of organization needed to manage permanent agricultural fields and systems of irrigation. Another centre of early crop irrigation using river water was in the Zaña Valley on the western slopes of the Peruvian Andes, where archeologists have unearthed a system of small-scale gravity canals that were being used at least 5,400 years ago, and probably 6,500 years ago.

Irrigated agriculture is arguably just as important today as it was to those early civilizations, and although several sources of fresh water are used to irrigate cropland – including groundwater, lakes, direct runoff, and various forms of wastewater – rivers remain by far the most important. The methods of storage (in reservoirs) and distribution (by canal) have not changed fundamentally since the earliest river irrigation schemes, with the exception of some contemporary projects' use of pumps to distribute water over greater distances. Nevertheless, many irrigation canals still harness the force of gravity. Half the world's large dams (defined as being 15 metres or higher) were built exclusively or primarily for irrigation, and about one-third of the world's irrigated cropland relies on reservoir water. In several countries, including such populous nations as India and China, more than 50% of arable land is irrigated by river water supplied from dams.

The knock-on effects of withdrawing water from a river to irrigate crops can be striking. In some cases, it may induce a complete transformation of river dimensions, pattern, and shape. One example of such 'river metamorphosis' comes from the western Great Plains of the USA, where rivers described by European Americans towards the end of the 19th century as wide, shallow, braided channels with only sparse vegetation along their banks have since been altered dramatically. The regulation of river flow for irrigated agriculture resulted in lower seasonal peak flows, higher base flows, and a change in regional water tables that promoted the establishment of trees along river banks. The combination of these changes to flow regime and bank resistance resulted in the rivers becoming narrow, sinuous channels flanked by dense forests within just a few decades.

Sadly, many irrigation schemes are not well managed and a number of environmental problems are frequently experienced as a result, both on-site and off-site. In many large networks of irrigation canals, less than half of the water diverted from a river or reservoir actually benefits crops. A lot of water seeps away through unlined canals or evaporates before reaching the fields. Some also runs off the fields or infiltrates through the soil, unused by plants, because farmers apply too much water or at the wrong time. Much of this water seeps back into nearby streams or joins underground aquifers, so can be used again, but the quality of water may deteriorate if it picks up salts, fertilizers, or pesticides. Excessive applications of irrigation water often result in rising water tables beneath fields, causing salinization and waterlogging. These processes reduce crop yields on irrigation schemes all over the world.

Many of these difficulties have plagued farmers in the Central Asian states of Turkmenistan and Uzbekistan where desert conditions mean that more than 90% of agriculture relies on irrigation from the Amu Darya and Syr Darya rivers. A rapid expansion of irrigation in Central Asia was initiated in the 1950s during the Soviet era, with some dramatic consequences. By the 1980s, the irrigated area had more than doubled to occupy about 7 million hectares. As a result, the annual inflow to the Aral Sea from the two rivers, the source of 90% of its water, had declined by an order of magnitude from about 55 cubic kilometres a year to some 5 cubic kilometres annually.

Unsurprisingly, the Aral Sea has become considerably smaller in consequence. In 1960, it was the fourth largest lake in the world, but since that time its surface area has more than halved, it has lost two-thirds of its volume, and its water level has dropped by more than 25 metres. In some areas, the Aral Sea's remaining waters are more than twice as salty as sea water in the open ocean. Most of the lake's native fish and other aquatic species have disappeared, unable to survive in the salt water, meaning an end to a once-major commercial fishing industry. Receding sea levels have also had local effects on climate, and the exposed sea bed has become a source of major dust storms that billow out over surrounding agricultural land up to several hundred kilometres from the Aral's coastline. This fine dust is laden with salts, adding to the problems of irrigated agriculture. It is also thought to have damaging impacts on human health.

Effects on fish

People have directly affected the biology of rivers over a very long period. In Europe, the common carp is found in the rivers of every country, but the fish is native only to the Danube and some of its tributaries. It was introduced to many European rivers by the Romans about 2,000 years ago after large numbers of legionnaires developed a taste for wild carp while stationed along the Danube – then the northern boundary of the Roman Empire – in the province of Pannonia.

This was how the common carp became the first species to be introduced into the Seine in France, for instance. It was followed in the Middle Ages by other species, including tench and rudd, which escaped from fish-farming ponds kept by noblemen and religious communities. In the late 19th century, further invaders (nase and pikeperch) arrived in the Seine from rivers further east via canals. These were followed at the end of the century by deliberate introductions of North American species: rainbow trout, black bass, pumpkinseed, and black bullhead.

Native fish began to disappear from the Seine in the 20th century as the construction of weirs and locks made it impossible for migratory species to reach their upstream spawning grounds. With the exception of the eel, all of the Seine's migratory species became extinct: sturgeon, salmon, sea lamprey, sea trout, European smelt, and shad. The original fish fauna of the Seine probably consisted of about 30 species. Today, the river has 46 species, but only 24 of them are native.

The catalogue of human impacts on the fish biology of the Seine is fairly typical of many rivers in the more economically developed parts of the world. Biological invasions generally are widely acknowledged to be one of the major threats to biodiversity across the world in rivers as well as other ecosystems. A study of the global patterns of freshwater fish invasions in more than 1,000 river basins covering more than 80% of the Earth's continental surface identified western and southern Europe as one of six global invasion hotspots, where non-native species represent more than one-quarter of the total number of species per basin. These hotspots also have the highest proportion of threatened fish species.

The human impact was found to be the most important determinant of this situation, particularly the level of economic activity, expressed by the gross domestic product, in a given river basin. The pattern can probably be explained in several ways. Regions that are economically prosperous are more prone to habitat disturbances (e.g. dams and reservoirs modifying river flows) that are known to assist the establishment of non-native species. High rates of economic activity are also likely to increase the chances of invading species arriving via aquaculture, sport fishing, and the ornamental trade. A higher demand for imported products associated with economic development also increases the likelihood of unintentional introductions occurring when imports are made.

Multiple human impacts also contribute to ecological change in rivers in poorer parts of the world, of course. Madagascar, like many islands, has a great many 'endemic' species (those found nowhere else) of all sorts, including fish, and its freshwater species are considered extremely vulnerable. Four of Madagascar's 64 endemic freshwater fish species are feared extinct, and another 38 are endangered due to three main pressures: habitat degradation caused by deforestation, overfishing, and interactions with exotic species.

Madagascar's widespread deforestation has contributed to the degradation of aquatic habitats in numerous ways. The loss of trees along river banks can result in changes in the species found in the river because fewer trees means a decline in plant matter and insects falling from them, items eaten by some fish. Fewer trees on river banks also results in less shade. More sunlight reaching the river results in warmer water and the enhanced growth of algae. A change in species can occur as fish that feed on falling food are edged out by those able to feed on algae. Deforestation also typically results in more runoff and more soil erosion. This sediment may cover spawning grounds, leading to lower reproduction rates. More sediment can also clog the gills of fish, causing them greater stress which, in combination with other pressures, can lead to their demise.

Overfishing of freshwater species is an intractable problem given the rising demand for fish from a rapidly increasing human population in Madagascar and the great logistical difficulties faced in enforcing any sort of environmental regulations. Exotic fish species introduced to the island include both aquacultural and ornamental species, and their impact on aquatic ecosystems has been profound. Some exotics have become naturalized, completely replacing native fish in the central highlands of Madagascar and becoming widespread in other parts of the island.

River regulation

Efforts to control the water level of rivers and the variability of river flows, to meet the demands of society, date back to the earliest civilizations. Today, rivers are regulated for many reasons, primarily to maintain an even flow for domestic, agricultural, and industrial needs, for hydroelectric power generation, for navigation, and to prevent flooding. The major methods employed in river regulation are the construction of large dams (see below), the building of run-of-river impoundments such as weirs and locks, and by channelization, a term that covers a range of river engineering works including widening, deepening, straightening, and the stabilization of banks.

13. The lower reaches of the West Lyn River flowing through Lynmouth in southwestern England were channelized after a devastating flood that killed 34 people in the village in 1952. The channel was widened and embankments built to increase its capacity

Some of the earliest scientific principles of channel regulation were established in Italy, where Leonardo da Vinci is credited with the invention of the pound lock using built-in vertical gates as a means of overcoming variations in river level. He produced the design at the end of the 15th century for a lock on the Naviglio Grande canal from Ticino to Milan, a development that acted as a considerable spur to inland navigation. Two hundred years later, the 'science of waters' was well established in northern Italy with the creation of a chair of 'hydrometry' at the University of Bologna in 1694 and publication of a host of books on river hydraulics. At this time it was advocated that regulation of braided rivers was best achieved by reducing them to a single channel, and by the end of the 19th century most braided rivers in western Europe had been regulated in this way.

Another important phase of river engineering that took place in Europe during the 19th century was the widespread straightening of channels and deepening of beds on major rivers. Significant work of this nature was conducted on the Seine in France and on the Sulina branch of the Danube delta, but one of the most dramatic schemes was implemented on the Tisza River, a tributary of the Danube that flows through Hungary. Regulation of the Tisza, designed to drain land for agriculture and reduce flooding on the Hungarian plain, involved cutting off more than 100 meanders, thus shortening the length of the river by nearly 400 kilometres.

The Yellow: 'China's Sorrow'

One of the most remarkable case histories of river management has unfolded over many centuries along the Yellow River, or Huanghe. It is the world's fourth longest river, although it is only number two in China after the Yangtze, and is also regarded as the world's siltiest, deriving its name from the 1.6 billion tonnes of fine yellow sediment it carries each year as it flows out of the Loess Plateau region and enters the North China Plain. The Yellow

River originates on the Tibetan Plateau and flows for more than 5,000 kilometres to the Bohai Sea, an inlet of the North Pacific Ocean. But it hasn't always done so. Like many rivers, the Yellow has changed its course over the years, albeit more frequently than most. In fact, over about the last 2,500 years, the Yellow River has averaged a major change in its course roughly every century. On some occasions, it has not flowed into the Bohai Sea, diverting into the Yellow Sea more than 300 kilometres to the south. For several hundred years, it didn't flow into the sea at all, but into a lake.

Every channel change has meant a major flood disaster on the densely populated plains of eastern China. Indeed, the Yellow's propensity to flood has earned it the nick-name of 'China's Sorrow'. One flood near the sizable city of Kaifeng in September 1642 drowned an estimated 340,000 people, leaving Kaifeng with a population of just 30,000 inhabitants. The Chinese began trying to prevent such floods more than 2,200 years ago by building up the Yellow's river banks with dykes or levées. At the beginning of the 21st century, levées lined the last 870 kilometres of the lower Yellow River to the sea. Constructing levées has probably saved many lives, but the banks have failed in numerous places over the years, still causing inundation on a catastrophic scale.

One of these levée failures, in 1938, was deliberate. During the war against the Japanese, the Chinese Nationalist government ordered its army to dynamite the levée at Huayuankou in an attempt to stop the advance of Japanese forces with an intentional flood. Although several thousand Japanese troops were drowned, the flood only delayed the enemy advance. The brunt of the disaster was borne by the local Chinese population. Eleven cities and more than 4,000 villages were inundated. In total, about 12 million people were affected, nearly 900,000 of them drowning. It was nine years before engineers repaired the levée at Huayuankou and the river resumed its course to the Bohai Sea.

Centuries of levée construction have had other effects. Most rivers in their lower courses deposit mud and silt, and the Yellow is no exception. However, because the river floods only rarely in its lower course, thanks to the levées, most of the material is deposited on the bed of the channel itself. Hence, the river channel has slowly gained height over the centuries, and the levées have had to be raised accordingly. Today, the bed of the lower reaches is on average some 5 metres higher than the land outside its dykes. At Kaifeng, the river bed is 13 metres higher than street level. The residents of Xinxiang go about their business no less than 20 metres below the adjacent Yellow River. The phenomenon is often referred to as a 'hanging river'.

Since the 1960s, a number of large dams and reservoirs have been built in the upper and middle reaches of the Yellow River. They are designed both to help control floods and to supply the 100 million people who rely on the river for their fresh water. The rising demands on the Yellow River's water have created a scarcity, to the extent that in the early 1990s the river failed to reach the sea on certain days. By 1997, there were 226 'no-flow' days, the dry point starting 700 kilometres inland on some occasions. Since then, the Chinese government has ensured for political reasons that the river always reaches the sea, albeit in small volumes. But the river now certainly delivers much less than a billion tonnes of sediment a year to the North Pacific. With so little water actually flowing in the hanging part of the river, the chances of a flood have decreased, but the possibility remains that a major flood further upstream will be too great for the dams to contain and the levées on the lower Yellow will once again be breached, with terrible consequences.

Dams

One of the most profound ways in which people alter rivers is by damming them. Obstructing a river and controlling its flow in this way brings about a raft of changes. A dam traps sediments and nutrients, alters the river's temperature and chemistry, and affects the processes of erosion and deposition by which the river sculpts the landscape. Dams create more uniform flow in rivers, usually by reducing peak flows and increasing minimum flows. Since the natural variability in flow is important for river ecosystems and their biodiversity, when dams even out flows the result is commonly fewer fish of fewer species.

Although dams have been built on rivers for thousands of years, the past 50 years or so has seen a marked escalation in the rate and scale of construction of dams all over the world, thanks to advances in earth-moving and concrete technology. At the beginning of the 21st century, there were about 800,000 dams worldwide, some towering more than 200 metres in height. Certain rivers have been intensively manipulated in this way. North America's River Columbia, for example, has, since the mid-19th century, become the site for no fewer than 80 dams. In some large river systems, the capacity of dams is sufficient to hold more than the entire annual average discharge of the river. The reservoirs behind dams on the Volta River in West Africa can store more than four times the river's annual average flow. Globally, the world's major reservoirs are thought to control about 15% of all runoff from the land. The volume of water trapped worldwide in reservoirs of all sizes is no less than five times the total global annual river flow, and this huge redistribution of water is thought to be responsible for a very small but measurable change in the orbital characteristics of the Earth.

The very first dams were constructed to control floods and to supply water for crop irrigation and domestic use. Modern dams still provide these services, plus a number of others, including hydroelectricity generation and industrial water supply. There is no doubt that many dam schemes have been very successful in achieving their objectives, and in many respects have made substantial contributions to the sustainable use of river resources. In Egypt, the Aswan High Dam has been perceived as a great symbol of economic advancement and national prestige since its completion in 1970. It generates about 20% of the country's electricity, and water held in its reservoir, Lake Nasser, has enabled irrigated agriculture to expand on to 5,000 square kilometres of new land. This is particularly important for a desert country with only a very small area suitable for cultivation. The creation of Lake Nasser has also given rise to a new fishing industry. The dam allows management of the highly seasonal variations in discharge, evening out the Nile's flow to protect against both floods and droughts. The stability of water levels in the river's course has also brought benefits for navigation and tourism.

Despite the success of many dams in achieving their main aims, however, their construction and the creation of an associated reservoir bring about significant environmental changes, many of which have proved to be detrimental. The precise nature and magnitude of changes vary greatly with the type of reservoir and the way it is operated, as well as according to the nature of the river basin affected. The most obvious impact of a new dam is the inundation of an area for its reservoir, with associated effects on hydrology, vegetation, wildlife, local climate, and even tectonic processes.

Reservoirs formed by river impoundment typically undergo significant variations in water quality during their first decade or so, before a new ecological balance is reached. Biological production can be high on initial impoundment, due to the release of organically bound elements from flooded vegetation and soils, but declines thereafter. One effect of the nutrient enrichment typical of new reservoirs, particularly in tropical and sub-tropical regions, is the blooming of toxic microscopic algae known as cyanobacteria. These cyanobacterial toxins are dangerous to humans and animals if consumed in sufficient quantities, causing a range of gastrointestinal and allergenic illnesses. Another biological consequence of large reservoirs is the rapid spread of waterweeds that cause hazards to navigation and a number of secondary

14. The Warragamba Dam near Sydney, Australia, is one of the world's largest domestic water supply dams. Its reservoir, which is 52 kilometres long, provides 80% of the water for about 4 million people in the Sydney region

impacts, notably the loss of water through evapotranspiration. A dramatic example occurred on the Brokopondo reservoir in Surinam during its first two years, where water hyacinth quickly proliferated to cover about half the lake's surface.

Some reservoirs are very large: Brokopondo covers an area of about 1,500 square kilometres, but Lake Volta, the reservoir behind the Akosombo Dam in Ghana, is more than five times as big, making it the world's largest man-made lake. The creation of such vast new bodies of water is thought to affect local climates. Following the establishment of Lake Volta, the peak rainfall season in central Ghana has shifted from October to July/August. Some particularly deep reservoirs can trigger earthquakes due to the stress on crustal rocks induced by huge volumes of water. Nurek Dam on the Vakhish River in central Tajikistan is one of the best-documented examples of a large dam causing seismic activity. This part of Central Asia is tectonically active anyway, but initial filling of the reservoir and each period of substantial increase in water level was mirrored by significant increases in earthquake frequency during the first decade of the dam's lifetime.

The building of a new dam means that any previous inhabitants of the area designated for the reservoir must be moved. The numbers of people involved can be very large and some of the biggest schemes in this respect have been in China. The Sanmen Gorge Project on the Yellow River involved resettling 300,000 people, and the Three Gorges Dam on the Yangtze River has displaced about 1.2 million people from 13 cities, 140 towns, and more than 1,000 villages. Governments usually offer compensation to people who are displaced by a new reservoir, but in many remote areas inhabitants do not possess formal ownership documents for the land they live on, a problem that can slow or actually prevent legal compensation.

Downstream of a reservoir, the hydrological regime of a river is modified. Discharge, velocity, water quality, and thermal characteristics are all affected, leading to changes in the channel and its landscape, plants, and animals, both on the river itself and in deltas, estuaries, and offshore. By slowing the flow of river water, a dam acts as a trap for sediment and hence reduces loads in the river downstream. As a result, the flow downstream of the dam is highly erosive. A relative lack of silt arriving at a river's delta can result in more coastal erosion and the intrusion of seawater that brings salt into delta ecosystems. Downstream changes in salinity due to construction of the Cahora Bassa Dam in Mozambique threaten mangrove forests at the mouth of the River Zambezi. One knock-on effect of this is a decline in prawns and shrimps, both of which breed in mangroves.

The most dramatic downstream effects have occurred on rivers dammed in several places. Construction of a series of dams in the 20th century on the Colorado River, one of the most intensively used waterways in the USA, has severely cut the river's naturally heavy sediment load, which had led the Spanish explorer Francisco Garces to name it (Rio Colorado is Spanish for 'red-coloured river'). Before 1930, the river carried more than 100 million tonnes of sediment suspended in its water each year to the delta of the Gulf of California, but it delivered neither sediment nor water to the sea from 1964, when Glen Canyon Dam was completed, to 1981, when Lake Powell behind the dam was filled to capacity for the first time. Since then, river water has reached the Gulf of California only irregularly, when discharges from dams allow. On average, the river now delivers an annual sediment load to the Gulf of California that is three orders of magnitude smaller than the pre-1930 average. The decline in fresh water and nutrients brought by the river to its estuary and the Gulf of California has had a huge impact on ecology. One study suggests that the lack of river-borne nutrients today may have resulted in a 96% decrease in the population of shellfish in the Colorado River Delta in Mexico.

The effects of dams on river ecology are numerous. Other important drivers of ecological impacts include changes in river temperature, the amount of dissolved oxygen carried, and the barrier effect of dams on the dispersal and migration of plants and animals. The dam-barrier effect on migratory fish and their access to spawning grounds has been recognized in Europe since medieval times. A statute introduced in Scotland in 1214 required all dams to be fitted with an opening and all barrier nets to be lifted every Saturday to allow salmon to pass. The problem certainly continues, however, sometimes with considerable economic implications. For example, a dramatic decline in catches of Caspian Sea sturgeon, the source of caviar, during the late 20th century was attributable primarily to the construction of several large hydroelectric dams on the River Volga and the consequent loss of spawning grounds.

Disruption to the movements of fish is one of several reasons for a recent movement in some countries to decommission dams. The small number of dams removed includes those that no longer serve a useful purpose, are too expensive to maintain, or have levels of environmental impact now deemed unacceptable. Most dams that have been removed or considered for removal are on rivers in the USA, but several European countries have also been involved in dam decommissioning. For example, two dams were destroyed and the debris cleared from tributaries of the Loire River in France in 1998 as part of a long-term government management scheme – the Plan Loire Grandeur Nature – for the river and its basin. A central aim of the scheme is to ensure the environmental protection of the Loire and to restore the river's salmon population. Removal of the Maisons-Rouges Dam on the River Vienne and the Saint Etienne de Vigan Dam on the River Allier was designed to restore access to salmon spawning grounds.

Land use

Rivers are intimately connected to the landscapes through which they flow, so it should come as no surprise to learn that any changes in a landscape inevitably affect its rivers. The way people use landscapes strongly influences rivers in numerous ways at multiple scales. Clearing natural forest vegetation to provide land for cultivation, for example, is well known to cause less interception of rainfall, less infiltration of water into the soil, less evapotranspiration, and more surface runoff, typically causing enhanced rates of soil erosion, in some cases by several orders of magnitude. Much of that soil finds its way into a river, causing associated changes in channel form and ecology. These types of alteration to rivers have been recorded all over the world, first occurring thousands of years ago in agricultural areas of the Mediterranean and China, and more recently elsewhere. Other forms of food production can also increase runoff and erosion. Grazing and trampling by livestock reduces vegetation cover and causes the compaction of soil, which reduces its infiltration capacity.

As rainwater passes over or through the soil in areas of intensive agriculture, it picks up residues from pesticides and fertilizers and transports them to rivers. In this way, agriculture has become a leading source of river pollution in certain parts of the world. Concentrations of nitrates and phosphates, derived from fertilizers, have risen notably in many rivers in Europe and North America since the 1950s and have led to a range of environmental, social, and economic problems encompassed under the term 'eutrophication' – the raising of biological productivity caused by nutrient enrichment. The growth of algae is the primary concern, leading to human health problems – and hence additional costs of water treatment for drinking – and effects on other river species. In slow-moving rivers, for example, the growth of algae reduces light penetration and depletes the oxygen content of the water, sometimes causing fish kills.

Of course, many of these effects can be controlled by conscious efforts to conserve soil and water on agricultural land. These sorts of measures are undertaken for all sorts of reasons, not least because losing soil and water from fields has an adverse effect on crop yields. Numerous studies undertaken in the Yellow River Basin in China have demonstrated the benefits of soil and water conservation measures, including tree planting and the construction of terraces, introduced in this area primarily to reduce sedimentation in the river's reservoirs. Discontinuing a land use that exacerbates runoff or sediment production is also likely to reduce these effects if the previous vegetation cover is re-established, but this does not always occur. Investigations in the central Andes of Peru found that where agricultural terraces had been abandoned, the rates of soil erosion increased because the environment was too dry for plants to grow on the terraces without attention from farmers.

Another form of land use that has similar effects is mining. In western Siberia, the sediment load of the Kolyma River more than doubled during the 1970s and 1980s due to widespread gold mining in the catchment disturbing vegetation and increasing erosion. Interestingly, records of the Kolyma's discharge over the same period showed no significant trend, indicating that runoff had remained the same. Many mining operations have also caused contamination in rivers. Waste rock and 'tailings' – the impurities left after a mineral is extracted from its ore – typically still contain metals which can be leached into soils and waterways. The accidental release of polluted water from a pond at the Aznalcóllar pyrite mine in southwest Spain in 1998 caused huge damage to birds, fish, and other aquatic species in the Guadiamar River and the Coto Doñana wetland. The water was acidic and contained arsenic, lead, and zinc at concentrations that were lethal for wildlife. Mining has long been associated with impacts on rivers. The Romans developed techniques of hydraulic mining, diverting large volumes of river water to break up and flush away soil and rock and expose minerals. The techniques were widely used to produce gold from alluvial deposits in northwest Spain.

The innumerable links between a river and human activities in its surrounding landscape, and consequently the importance of managing an entire basin, have been recognized for centuries. In Japan, for example, government regulation of timber harvesting along mountain streams in order to maintain channel stability dates back 1,200 years. Similarly, the traditional Hawaiian systems of *ahupua'a* involved managing drainage basins as an integrated whole to safeguard food production from agriculture and fish ponds. Upland forests were protected by taboo in order to supply rivers with nutrients for downstream fields and fish ponds. In modern parlance, the approach is embodied in 'catchment management plans' which in the countries of the European Union have become mandatory for all major river basins.

The Mississippi

The Mississippi River which, together with the Missouri River, drains two-thirds of the continental USA, has been significantly modified by numerous human activities over the last 200 years or so. A rapid rise in river traffic dating from the beginning of steam boats in the early 1800s spurred the large-scale felling of forests to fire the boats' boilers, and the loss of trees in turn destabilized river banks and contributed to unpredictable migration of the channel. Deforestation and the expansion of commercial agriculture in the Mississippi Basin also resulted in more soil erosion and more sediment reaching the river. Sandbars, a menace to navigation, were one result. As settlements expanded on to low-lying river banks, the Mississippi's floods became a greater danger.

Attempts to manage these problems on the Mississippi in a systematic way began in the 19th century and continue today. Throughout the 1800s, the US Army Corps of Engineers cleared rock and made the channel deeper on particular stretches of the river in an effort to assist navigation. A major programme of river engineering was initiated after a disastrous flood in 1927 in the Lower Mississippi Valley which cost more than 200 lives and displaced over 600,000 people. The Mississippi River and

15. Maintaining the Mississippi's role as an important transport route has been the motivation for many human impacts on the river. This barge is near Baton Rouge, Louisiana, where the channel is about 700 metres wide

Tributaries Project was designed to control flooding and improve navigation in several ways, and one of these was to straighten the channel by eliminating meanders. Artificially creating a meander cut-off shortens the course of the river, so increasing its gradient and speed of flow. In this way, the water erodes and deepens the channel, thereby increasing its flood capacity. The huge scale of this operation was reflected in a dramatic shortening of the Mississippi. In 1929, a boat sailing between Memphis, Tennessee, and Red River Landing in Louisiana travelled 885 kilometres, but by 1942 the same journey had been shortened by 274 kilometres – some 30% – thanks to the series of cut-offs.

Further protection from floods is provided by nearly 3,500 kilometres of levées and floodwalls along the Mississippi River itself and along some of its major tributaries, but despite these huge efforts the Mississippi is still prone to flooding. The 1993 flood on the river's upper reaches ranks as one of the worst natural disasters in US history, destroying or seriously damaging more than 40,000 buildings. Heavy rain caused the river to breach the levées in more than 1,000 places, and in many locations flooding was prolonged because levées prevented the return of water to the channel once the peak had passed. It also seems very likely that efforts to manage the flood hazard on the Mississippi have contributed to an increased risk of damage from tropical storms on the Gulf of Mexico coast. The levées built along the river have contributed to the loss of coastal wetlands, starving them of sediment and fresh water, thereby reducing their dampening effect on storm surge levels. This probably enhanced the damage from Hurricane Katrina which struck the city of New Orleans in 2005.

Urban rivers

Cities have had numerous impacts on rivers, starting with the rise of the first urban civilizations which emerged on the floodplains of large rivers in several parts of the world a few thousand years ago (see Chapter 3). Archaeological excavations at Harappa and Mohenjo Daro in the Indus Valley have revealed ceramic pipes designed to supply water and brick conduits under the streets for drainage that are thought to have been in operation 5,000 years ago. The Romans are also well known for their sophisticated water-supply systems. Water was brought to Ancient Rome from distant streams and springs via nine major aqueducts. Some of these were more than 60 kilometres in length and involved tunnelling through difficult hillsides with vertical shafts dug for inspection and cleaning.

Large amounts of water were involved in these early municipal systems, but they were ultimately limited in the volume of water managed by the force of gravity. Water could be transferred from one place to another only as long as the direction was down a slope. Modern civilization has hugely increased its ability to move water by using energy to pump water. In the southwestern USA, for example, water from the Colorado River is pumped nearly 500 kilometres across the Mojave Desert to large cities on the west coast of California, including Los Angeles and San Diego.

The growth and development of urban areas – the process of urbanization – is frequently associated with such changes to river systems, some deliberate, others inadvertent. Deliberate manipulation of rivers can be on a significant scale. In Japan, for instance, the city of Tokyo began to develop rapidly after the mid-17th century realization of the Tone River Easterly Diversion Project, a grand scheme that took more than 50 years to complete and involved diverting the Tone River more than 100 kilometres to the east to prevent flooding of the nascent city. The early stages of urban development typically result in a number of other, more subtle effects on rivers. Trees and other vegetation are removed prior to construction which results in less interception of rainfall and less transpiration, both leading to a greater flow of water and more erosion of the bare surfaces, often leading to sedimentation within river channels. In places where scientists have monitored soil erosion from construction sites, the yield of sediment has been up to 100 times greater than under natural conditions. In one extreme case, a rate of more than 600,000 tonnes of soil a year was measured from an abandoned construction site in Kuala Lumpur in Malaysia, about 20,000 times the natural erosion rate.

Other impacts in the early stages of urban development stem from growing numbers of people drawing water directly from a river or drilling wells which can indirectly affect river hydrology by lowering the water table. Rivers also provide a ready source of modern construction materials, and the excavation of sand and gravel can have significant impacts on the geometry and ecology of a river.

One of the most important effects cities have on rivers is the way in which urbanization affects flood runoff. Large areas of cities are typically impermeable, being covered by concrete, stone, tarmac, and bitumen. This tends to increase the amount of runoff produced in urban areas, an effect exacerbated by networks of storm drains and sewers. This water carries relatively little sediment (again, because soil surfaces have been covered by impermeable materials), so when it reaches a river channel it typically causes erosion and widening. Larger and more frequent floods are another outcome of the increase in runoff generated by urban areas.

Contamination of river water has always been an issue in large urban areas, but particularly serious water pollution problems occurred with the growth of cities during the Industrial Revolution thanks to large volumes of domestic sewage and industrial effluents. Water quality in the River Thames at London declined throughout the first half of the 19th century as the city's population grew, along with a rapid increase in the use of the flushing water closet. Untreated sewage flowed directly into the river, along with liquid wastes from a growing number of factories, slaughter houses, tanneries, and other industries located on the Thames.

Organic liquid wastes such as sewage and effluent from industries that process agricultural products can be broken down by bacteria and other micro-organisms in the presence of oxygen. An overload of such organic wastes leads to decreasing levels of dissolved oxygen in a river, so that fish and aquatic plant life suffer and may eventually die. By 1849, fish had disappeared from the tidal Thames, which included the entire length of the river in London. At this time, river water was still being abstracted for public consumption and water-related diseases were rife: five cholera epidemics occurred in London between 1830 and 1871. During the long, dry summer of 1858, the so-called Year of the Great Stink, the Houses of Parliament had to be abandoned on some days because of the terrible stench from the river.

Such a direct impact on the nation's politicians produced some positive action, and conditions in the Thames had improved by the 1890s with the introduction of sewage treatment plants. During the first half of the 20th century, however, sewage treatment and storage did not keep pace with London's growing population, and the oxygen content of the river reached zero 20 kilometres downstream of London Bridge during many summers. Water quality gradually improved after 1950 with tighter controls on effluent and improved treatment facilities. By the 1970s, the river's water was widely regarded as satisfactory, and in 1974 much publicity accompanied the landing of the first salmon caught in the Thames since 1833.

A similar story can be told for rivers flowing through major cities in many parts of the industrialized world: a rapid increase in pollution that accompanies industrialization and population growth leading, in time, to the implementation of pollution controls and recovery to a tolerable environmental quality. Some of the early 21st century's most polluted urban rivers are in the rapidly industrializing parts of Asia. They include the Buriganga River in Dhaka, Bangladesh; the Marilao River in Metro Manila, in the Philippines; the Citarum River near Jakarta, Indonesia; and the Yangtze River which flows through numerous cities in China.

Controlling river blindness

Flooding is the most widespread hazard to human society associated with rivers, but in certain parts of the world a disease named onchocerciasis, or river blindness, is a more enduring concern. The disease is caused by a parasitical worm that is transmitted among humans by the bites of small black flies which breed in rapid-flowing rivers and streams. Once inside the human body, the worms form disfiguring nodules on the skin and their tiny larvae move, causing blindness if they reach the eye. The World Health Organization estimates that more than 17 million people are infected worldwide, some half a million of whom are visually impaired.

River blindness occurs in parts of tropical Africa, Latin America, and the Arabian peninsula. The presence of the parasite in Latin America is almost certainly a result of infected people moving to the Americas, probably as part of the slave trade. The highest prevalence and the most serious manifestations of the disease still occur in West Africa despite the significant success of a huge programme initiated in the early 1970s to control the disease. The Onchocerciasis Control Programme in West Africa focused on controlling the black fly that transmits the disease by spraying vast stretches of West African rivers with insecticide. At the peak of the programme, this involved more than 50,000 kilometres of river over an area of more than a million square kilometres in 11 countries. Spraying was frequent, almost weekly for 10 to 12 months each year, in some cases over a period of 20 years. The idea was to stop transmission of the parasite for the duration of the life span of the worm in humans, considered to be more than a decade.

This ambitious programme is thought to have protected some 40 million people in West Africa from river blindness and opened up 250,000 square kilometres of land in previously infected river valleys to resettlement and cultivation. Monitoring of other insects, and fish, in the treated rivers indicated few deleterious effects, and the current view of river ecologists is that permanent damage to other creatures in these rivers is unlikely.

Global warming

The human-induced warming of the global climate has issued in a new era of society's influence on rivers. An overall increase in temperature will melt snow and ice and translate into a greater loss of moisture from soils due to higher evaporation and transpiration from plants. River flows will also be affected by changes in precipitation amounts, the intensity and duration of storms, their timing, and the type of precipitation involved. Climatologists agree that extreme weather events (examples include tropical cyclones, droughts, heat waves, and heavy rainstorms) are likely to become more frequent, more widespread, and/or more intense in many parts of the world as the 21st century progresses. All will inevitably result in changes to rivers. Less direct, but potentially no less significant, changes will also occur due to the ways in which plant communities respond to climatic warming. Societies too can be expected to increase their influence on some rivers in response to other aspects of climate change; expanding irrigation systems, for example, in regions subject to more droughts.

Detecting the impact of global warming on rivers is by no means always straightforward because of the difficulties of separating the effect of climate change from the natural variability of many fluvial characteristics and the need to take account of possible alternative causes of change, such as alterations to land use and other human activities. Nonetheless, the influence of global warming has already been identified in some recent modifications to fluvial systems. Work on a number of the world's large drainage basins has established a significant rising trend in the risk of great floods (those with a return period of 100 years) in the 20th century. Warmer air temperatures are also having a predictable effect on glaciers – melting and retreat – in many parts of the world. Glaciers are receding particularly fast in the Himalaya and parts of Tibet, generating worries about long-term water supplies for hundreds of millions of people in India, Bangladesh, Nepal, and China who rely on rivers fed by glacial meltwater.

Ice cover has been in general decline since the mid-19th century on most rivers in North America and Eurasia as gradual warming has meant freeze-up dates have been occurring later and break-up dates arriving earlier. In the case of the lower Don River in Russia, the length of the ice season has been reduced by a whole month in about 100 years. Records for the Tornionjoki River in Finland stretch back to 1692 and show a long-term trend towards earlier break-up dates throughout the entire period. This tendency is not universal, however. Rivers in central and eastern Siberia display significant trends in the opposite direction: towards longer periods with ice cover due to earlier freeze-up dates and later break-up dates.

Northern hemisphere rivers that flow into the Arctic Ocean have been delivering more water in line with longer ice-free periods, combined with an increase in precipitation. More fresh water in the Arctic could slow down or shut off the so-called 'thermohaline circulation', an oceanic current conveyor belt which transports large amounts of warm water to the North Atlantic region. This circulation is driven by differences in the density of sea water, controlled by temperature and salinity, so more fresh water could counteract the flow. The thermohaline circulation helps to regulate the climate of northern Europe, maintaining temperatures that are higher than would be otherwise expected given the latitude.

Conversely, a number of other rivers have seen declines in the amount of water they carry each year since the mid-20th century. Several of the major rivers with dwindling flows serve large populations, sparking further concerns about future water supplies. These rivers include the Yellow River in northern China, the Ganges in India, the Niger in West Africa, and the Colorado in North America.

Drought is thought to be the greatest agent of change associated with global warming in the Amazon Basin. Many computer-based models of future climate in the region indicate a reduction of dry-season rainfall, the effects of which will be exacerbated by rising air temperatures. This increased probability of drought will have all sorts of knock-on effects for the forest ecosystems and the rivers running through them, including a greater likelihood of fire. The consequences for local people, wildlife, and the rivers themselves are expected to be serious.

In Europe, the discharge of the Rhine is expected to become more seasonal because of global warming. Estimates generated by computer models indicate that by 2050, the average flow in summer will decrease by up to 45% and the average winter flow will increase by up to 30%. Less water in the Rhine during the summer months is related mainly to predicted decreases in precipitation and increases in evapotranspiration. Greater flows in winter will be caused by a combination of more precipitation, less snow storage, and increased early melting. The hazards posed by winter floods on the Rhine will certainly increase in consequence. Greater seasonality in the river's flow will also have numerous repercussions for the ecology of the Rhine.

River restoration

The numerous ways in which human activities have influenced rivers, both purposefully and indirectly, are complemented in many countries by efforts to reverse some of the earlier effects of human action: so-called 'river restoration'. Attempts to improve conditions in rivers are not new in themselves, as evidenced in the clean-up of the Thames in London cited earlier in this chapter, for instance, but the widespread adoption of restoration, rehabilitation, and mitigation measures has been recognized as a distinctive phase of river management in the late 20th and early 21st centuries. Restoration projects usually involve efforts to repair damage to rivers, typically in an attempt to better meet societies' needs and expectations for natural, ecologically healthy waterways.

Returning a river to its 'natural' or 'original' condition is usually fraught with difficulty, however. Theoretically, at least, this can be based on an understanding of historical conditions along a river before human effects, or on conditions along a similar but less affected reference river. In practice, however, an appropriate reference river may not exist, or conditions in a basin (such as climate or vegetation) may have changed since the period selected for the historical baseline. Indeed, rivers change under all sorts of natural circumstances, and determining which changes are natural and which are due to human pressures is not always straightforward. Further, although it may be possible to determine which human impacts are undesirable, preventing them entirely may be more complicated.

These and other constraints mean that re-establishing conditions that might have existed prior to human settlement of the landscape is virtually impossible. It is more appropriate to restore rivers that are self-sustaining and integrated into the surrounding landscape and are, therefore, generally closer to a more natural state. Hence, for example, the Plan Loire Grandeur Nature for the Loire River in France, one of the largest river restoration programmes undertaken anywhere, aims to ensure the conservation of typical Loire Valley ecosystems (including peatlands, gorges, alluvial forests, and oxbow lakes) on model sites and to maintain their ecological functions. Part and parcel of this effort is the re-establishment of iconic river species such as the beaver and salmon.

Even when the objectives of river restoration programmes are clear, in most cases they will still have to be balanced against other demands put on rivers. Some of these demands may be conflicting. For example, some conservationists argue that river regulation and environmental conservation are intrinsically incompatible since regulation modifies the natural environment in which original wildlife communities became established. Indeed, in certain cases, the ecological requirements of organisms are destroyed or modified beyond the limits of adaptations and the organisms are unable to survive. River management is no different from any other natural environmental management issue in that it involves compromises, and in a world where the growth of populations and economies appears to be inexorable, not to mention the all-embracing effects of human-induced climate change, these compromises are likely to become more and more delicate.

Tamed rivers

Epilogue

Virtually every reader of this book will have some sort of relationship with a river, or perhaps more than one. It may involve living on a floodplain or benefiting directly from a river's flow, maybe as an angler or via a system of plumbing. The number of ways in which rivers impinge on human society is great, and there are few places on Earth where rivers do not exert an influence, be it direct or indirect, current or historical.

The aim of this book has been to celebrate rivers in all their diversity. Bountiful yet capricious, rivers represent different things to different people, sometimes contradictory, at others complementary. They form vital components to innumerable ecosystems, and nourish both town and country. That nourishment has been spiritual as much as literal. Rivers are worshipped and revered, respected and feared. From raging torrents to babbling brooks, their waters have fuelled the thoughts of artists, scientists, philosophers, and generals. In a very real sense, much of human history has taken place on the banks of rivers.

The ancient Greek philosopher Heraclitus of Ephesus asserted that 'you cannot step twice into the same river'. All rivers are inherently dynamic. A meandering channel can abruptly become braided, or a trickle burst out of its banks to inundate a plain. So too the life they sustain, from mountain peaks to muddy deltas, on timescales ranging from the nymphal life of a mayfly to the extinction of the Yangtze river dolphin. Humankind's use and abuse of rivers has been equally diverse and vibrant, ranging from development of the earliest river boats to the choking of waterways with industrial effluent.

River channels occupy just a tiny fraction of the land surface, but their influence is out of all proportion to this immediate footprint. No matter how you may perceive rivers, all must acknowledge the wide and eclectic menu of river-based themes. Together, they reflect both the natural and social history of our planet.

Further reading

P. Ackroyd, *Thames: Sacred River* (London: Chatto & Windus, 2007).

K. J. Avery and F. Kelly, *Hudson River School Visions: The Landscapes of Sanford R. Gifford* (New York: Metropolitan Museum of Art Publications, 2003).

B. K. Belton, *Orinoco Flow: Culture, Narrative, and the Political Economy of Information* (Lanham: Scarecrow Press, 2003).

A. C. Benke and C. E. Cushing (eds.), *Rivers of North America* (Amsterdam: Academic Press, 2005).

T. M. Berra, *Freshwater Fish Distribution* (Chicago: University of Chicago Press, 2001).

I. C. Campbell (ed.), *The Mekong: Biophysical Environment of an International River Basin* (Amsterdam: Elsevier Press, 2010).

J. Cao, *China Along the Yellow River: Reflections on Rural Society* (Abingdon: Routledge Curzon, 2005).

M. Cioc, *The Rhine: An Eco-Biography, 1815–2000* (Seattle: University of Washington Press, 2002).

F. S. Colwell, *Rivermen: A Romantic Iconography of the River and the Source* (Montreal: McGill-Queen's University Press, 1989).

N. Compton, *The Battle for the Buffalo River: A Twentieth-Century Conservation Crisis in the Ozarks* (Fayetteville: University of Arkansas Press, 1992).

J. Conrad, *Heart of Darkness* (London: Penguin Classics, 1973).

S. Darby and D. Sear, *River Restoration: Managing the Uncertainty in Restoring Physical Habitat* (Chichester: Wiley, 2008).

L. de Waal, P. M. Wade, and A. Large, *Rehabilitation of Rivers: Principles and Implementation* (Chichester: Wiley, 1998).

D. Dudgeon, *Tropical Stream Ecology* (Amsterdam: Academic Press, 2008).

M. D. Evenden, *Fish Versus Power: An Environmental History of the Fraser River* (New York: Cambridge University Press, 2004).

A. Feldhaus, *Water and Womanhood: Religious Meanings of Rivers in Maharashtra* (New York: Oxford University Press, 1995).

P. Fradkin, *A River No More: The Colorado River and the West*, 2nd edn. (Berkeley: University of California Press, 1996).

P. S. Giller and B. Malmqvist, *The Biology of Streams and Rivers* (Oxford: Oxford University Press, 1998).

A. L. Godinho, B. Kynard, and H. P. Godinho, *Life in a Brazilian Floodplain River: Migration, Spawning, and Management of São Francisco River Fishes* (Saarbrücken: LAP Lambert Academic Publishing, 2010).

W. Grady (ed.), *Dark Waters Dancing to a Breeze: A Literary Companion to Rivers and Lakes* (Vancouver: Greystone Books, 2007).

S. de Gramont, *The Strong Brown God: The Story of the Niger River* (Boston: Houghton Mifflin, 1975).

A. Gupta (ed.), *Large Rivers: Geomorphology and Management* (Chichester: Wiley, 2008).

J. Harding, P. Mosley, C. Pearson, and B. Sorrell, *Freshwaters of New Zealand* (Wellington: New Zealand Hydrological Society/New Zealand Limnological Society, 2004).

S. M. Haslam, *The Riverscape and the River* (Cambridge: Cambridge University Press, 2008).

J. Hemming, *Tree of Rivers: The Story of the Amazon* (London: Thames & Hudson, 2008).

J. F. Hornig (ed.), *Social and Environmental Impacts of the James Bay Hydroelectric Project* (Montreal: McGill-Queen's University Press, 1999).

P. J. Jones, *Reading Rivers in Roman Literature and Culture* (Lanham, MD: Lexington Books, 2005).

R. Kingsford (ed.), *Ecology of Desert Rivers* (Cambridge: Cambridge University Press, 2006).

L. B. Leopold, *Water, Rivers and Creeks* (Sausalito, CA: University Science Books, 1997).

M. C. Lucas and E. Baras, *Migration of Freshwater Fishes* (Oxford: Blackwell Science, 2001).

C. Mauch and T. Zeller (eds.), *Rivers in History: Perspectives on Waterways in Europe and North America* (Pittsburgh: University of Pittsburgh Press, 2008).

A. Meadows and P. S. Meadows (eds.), *The Indus River: Biodiversity, Resources, Humankind* (Karachi: Oxford University Press, 1999).

M. Meybeck, G. de Marsilly, and E. Fustec (eds.), *La Seine en son Bassin: Fonctionnement écologique d'un système fluvial anthropisé* (Paris: Elsevier, 1998).

S. Mithen and E. Black, *Water, Life and Civilisation: Climate, Environment and Society in the Jordan Valley* (Cambridge: Cambridge University Press, 2011).

C. Morris, *The Big Muddy: An Environmental History of the Mississippi and Its Peoples, from Hernando de Soto to Hurricane Katrina* (New York: Oxford University Press, 2010).

P. K. Parua, *The Ganga: Water Use in the Indian Subcontinent* (Dordrecht: Springer, 2010).

A. Poiani (ed.), *Floods in an Arid Continent, Advances in Ecological Research* 39 (San Diego: Academic Press, 2006).

J. D. Priscoli and A.T. Wolf, *Managing and Transforming Water Conflicts* (Cambridge: Cambridge University Press, 2009).

R. Randolph Ashton, *A Celebration of Salmon Rivers* (Mechanicsburg: Stackpole Books, 2007).

C. W. Sadoff, D. Whittington, and D. Grey, *Africa's International Rivers: An Economic Perspective* (Washington, DC: World Bank, 2002).

R. Said, *The River Nile: Geology, Hydrology and Utilization* (Oxford: Elsevier Science, 1993).

S. M. A. Salman and K. Uprety, *Conflict and Cooperation on South Asia's International Rivers: A Legal Perspective* (Washington, DC: World Bank, 2002).

S. A. Schumm, *River Variability and Complexity* (Cambridge: Cambridge University Press, 2005).

T. Scudder, *The Future of Large Dams* (London: Earthscan, 2006).

P. Sinclair, *The Murray: A River and its People* (Carlton South: Melbourne University Press, 2001).

D. E. Spritzer, *Waters of Wealth: The Story of the Kootenai River and Libby Dam* (Boulder: Pruett, 1979).

K. Tockner, U. Uehlinger, and C. T. Robinson (eds.), *Rivers of Europe* (Amsterdam: Academic Press, 2009).

S. Turvey, *Witness to Extinction: How We Failed to Save the Yangtze River Dolphin* (Oxford: Oxford University Press, 2008).

E. E. Wohl, *A World of Rivers: Environmental Change on Ten of the World's Great Rivers* (Chicago: University of Chicago Press, 2010).

"牛津通识读本"已出书目

古典哲学的趣味	福柯	地球
人生的意义	缤纷的语言学	记忆
文学理论入门	达达和超现实主义	法律
大众经济学	佛学概论	中国文学
历史之源	维特根斯坦与哲学	托克维尔
设计,无处不在	科学哲学	休谟
生活中的心理学	印度哲学祛魅	分子
政治的历史与边界	克尔凯郭尔	法国大革命
哲学的思与惑	科学革命	民族主义
资本主义	广告	科幻作品
美国总统制	数学	罗素
海德格尔	叔本华	美国政党与选举
我们时代的伦理学	笛卡尔	美国最高法院
卡夫卡是谁	基督教神学	纪录片
考古学的过去与未来	犹太人与犹太教	大萧条与罗斯福新政
天文学简史	现代日本	领导力
社会学的意识	罗兰·巴特	无神论
康德	马基雅维里	罗马共和国
尼采	全球经济史	美国国会
亚里士多德的世界	进化	民主
西方艺术新论	性存在	英格兰文学
全球化面面观	量子理论	现代主义
简明逻辑学	牛顿新传	网络
法哲学:价值与事实	国际移民	自闭症
政治哲学与幸福根基	哈贝马斯	德里达
选择理论	医学伦理	浪漫主义
后殖民主义与世界格局	黑格尔	批判理论